Royal Navy Aircraft

Royal Navy Aircraft

since 1945

Ray Williams

Naval Institute Press

This book is dedicated to the late
Commander (A) R. I. Gilchrist MBE VRD RNVR
and the officers and men of Nos. 1831 and 1841
RNVR Squadrons, Northern Air Division
at RNAS Stretton 1947-1957.

In memory of my mother and father,
May Victoria and John Williams.

First published 1989 in the UK
by Airlife Publishing Ltd.

Published and distributed in the United States of
America and Canada by the Naval Institute Press,
Annapolis, Maryland 21402.

Library of Congress Catalog Card No. 89-60742

ISBN 0-87021-996-0

Contents

Acknowledgements

The Author would like to record his appreciation of the assistance given by the following people during the preparation of this book:

G. Atkins, J. Barnard, S. Holliman and J. Titley of Rolls Royce; E. Barker, D. Cott, G. Ferguson, Mrs C. Johnson and Mrs C. Lang of British Aerospace; R. Barker and T. Calloway of the RAF Museum; Lt. G. N. Bowker RN; A. Bye of Westland Helicopters; A. Byrom; C.P.O. M. Debenham, C.P.O. B. A. Savory and C.P.O. B. Tredwill of FONAC; Cdr. P. A. Fish RN; R. Fox of Flight Refuelling; Lt. Cdr. D. R. George AFC RN; T. Goyer of Shorts; E. A. Harlin; J. M. Hepworth; H. Holmes; A. E. Hughes; G. Hughes; P. Jarrett; G. A. Jenks; Lt. Cdr. M. S. Lay RN; L. Lovell and G. Mottram of the F.A.A. Museum; L/A (Phot) P. Ball; B. J. Lowe; Cdr. K. D. Mackenzie RN; D. E. Monk; E. B. Morgan; Mrs J. M. Mote; E. Myall; J. D. Oughton; R. M. Rayner; Cdr. H. W. Rickard RN; R. T. Riding; J. A. Simpson; Cdr. T. J. K. Sloane RN; A. Stott; M. Stroud; R. C. Sturtivant; A. Thomas; Cdr. S. C. Thornewill DSC RN; C.P.O. (Phot) T. J. Tierney; Second Officer H. Tuppen WRNS; B. Wexham of Vickers; and N. G. Williams.

Introduction

The Royal Navy's involvement with aviation started in 1908 when the Admiralty gave approval for the construction of the No. 1 Rigid Naval Airship. This first venture unfortunately ended in disaster in 1911 when the aptly named Mayfly broke its back while it was being taken out of its shed in a strong wind, and was subsequently scrapped without ever having flown. Despite this setback the Admiralty was firmly committed to aviation. In February 1911 Mr. Francis Maclean, a member of the Royal Aero Club, had offered two of his aircraft, and the facilities at the aerodrome at Eastchurch to train four Royal Navy officers to fly. These first four pilots included both Lt. Cdr. C. R. Samson and Cdr. R. Gregory. That same year, the first take-off from water by a seaplane was achieved in November by Cdr. O. Schwann flying an Avro Biplane. Two months later Samson flew the Short S.27 from a platform on the deck of HMS *Africa* and on 9 May 1912 Gregory, flying the same Short, made the first British flight from a ship that was under way. These latter trials were carried out from HMS *Hibernia,* which had been fitted with a take-off ramp on the forward deck.

The Royal Flying Corps (RFC) was formed on 13 May 1912, with separate Naval and Military (Army) wings, sharing common training facilities at the Central Flying School. On 1 July 1914 the Naval Wing achieved complete independence from the RFC and became the Royal Naval Air Service (RNAS). At the start of the First World War the RNAS comprised 71 aircraft, seven airships and some 800 officers and men. However, during the war there was a massive increase and in just four years the RNAS grew to 2949 aircraft, 103 airships and over 67000 officers and men. Early in the war several seaplane carriers entered service, primarily to provide aircraft for reconnaissance and spotting duties. The first landplane to land on a ship was a Sopwith Pup flown by Sqn. Cdr. E. H. Dunning, which landed on HMS *Furious.* The technique was very hazardous and required considerable skill to fly alongside the ship and then to sideslip the aircraft round the front of the superstructure and funnel before touching down on the flat deck at the front of the ship. Before the end of the First World War the first true aircraft carrier, HMS *Argus,* was under construction, being the first to have an unobstructed flight deck.

The Royal Air Force (RAF) was formed from the RFC and the RNAS, much to the chagrin of the Admiralty, which for the next 20 years ran a campaign to regain control of naval aviation and only succeeded in doing so in May 1939 shortly before the outbreak of World War Two, although they had some success in 1923 when a committee of enquiry recommended that all observers and 70 per cent of Fleet Air Arm pilots should be Naval officers. In 1924 the title "Fleet Air Arm" had come into use, to cover naval aviation, but in the view of the Royal Navy the full title was considered to be the Fleet Air Arm of the RAF. In fact the feeling against the title "Fleet Air Arm" was so intense in official Royal Navy circles that shortly after the Second World War the title was abolished and replaced by "Royal Naval Air Service". However, by this time "Fleet Air Arm" was in regular usage and refused to go away, so in 1953 the Admiralty bowed to the inevitable and reverted to "Fleet Air Arm".

Aircraft carriers made some major steps forward between the wars, and in addition to HMS *Furious,* which had been modified to introduce a through deck, and HMS *Argus* from the First World War, a further six had entered service by the start of World War Two. The last of these was HMS *Ark Royal,* which had been commissioned in November 1938. During 1937 a major step forward was taken when four armoured carriers,

HMS *Argus,* commissioned in September 1918, was the world's first flush-decked aircraft carrier, and continued in service until 1944. (RAF Museum)

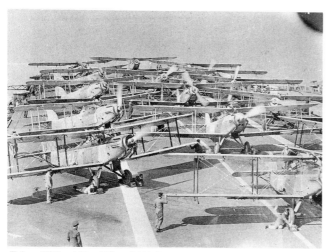

Fairey IIIFs of Nos 820 and 821 Squadrons preparing for take-off on HMS *Courageous* in May 1933.

Illustrious, Victorious, Formidable and *Indomitable,* were laid down, followed in 1939 by a further two, *Implacable* and *Indefatigable.* During the 1930s progress had been made with arrested landings and catapult launches, and this led to a system of transverse arrester wires fitted to friction brake-drums and using hydraulically operated resetting gear, a system that was still in use when the last of the conventional carriers left the scene in 1978. The catapult, or accelerator as it was known initially, was a hydro-pneumatic system which was installed under the carrier deck and launched the aircraft by means of a towing bridle operating through a slot in the deck. This system remained unchanged until the introduction of the steam catapult in 1954.

During the inter-war period, development of naval aircraft had progressed relatively slowly and, apparently because of the limited financial resources allocated to the Fleet Air Arm, only small quantities of aircraft were ordered. At the end of the First World War the RNAS was using several different types on carriers, including Sopwith Pups, 1½ Strutters and Camels as fighters, and the Sopwith Cuckoo, a torpedo bomber. These types did not survive long after the war and were replaced by a new range of aircraft designed specifically to meet requirements for carrier-borne aircraft. Two Fairey aircraft, the IIID (developed later into the IIIF) and the Flycatcher, entered service in 1923 and became, for the next ten years, the mainstay of the Fleet Air Arm in the spotter-reconnaissance and fleet fighter roles respectively. This period was also to see the introduction of several of what can only be described as very ugly aircraft, including the Blackburn Dart torpedo bomber and the Blackburn Blackburn and Avro Bison spotter-reconnaissance aircraft. The aircraft operated in flights until 3 April 1933 when there was a re-organisation of the Fleet Air Arm structure and a system of squadrons was introduced starting with No. 800 Squadron.

Replacement of the Flycatcher by the Hawker Nimrod, a navalised Fury, had started in 1932 and at about the same time the two-seat Hawker Osprey, a navalised Hart, had been introduced as a fighter-reconnaissance aircraft. The following year the Fairey 111Fs started to be replaced by the Fairey Seal. By the outbreak of the Second World War the inadequacies of the Fleet Air Arm's aircraft was apparent, with the latest types being Gloster Sea Gladiators, Blackburn Sharks, Fairey Swordfish and the only monoplane, the

Blackburn Skua, which was only capable of 225mph: and this was at a time when the RAF was re-equipping with Hurricanes and Spitfires. In October 1924 the Fleet Air Arm had 144 aircraft operated by 24 flights and in September 1939 this had only increased to 225 aircraft with 20 squadrons operating from carriers plus 115 aircraft allocated to the catapult flights; which by then had been introduced on all the major Royal Navy warships.

Although the new armoured fleet carriers that were starting to enter service early in the war were in advance of any of their contemporaries, the aircraft appeared to have been almost totally neglected, with the Fleet Air Arm entering the war with aircraft that were totally outclassed by their land based contemporaries. The first major step forward by the Fleet Air Arm with regard to aircraft was the purchase of Grumman Wildcats, which they renamed the Martlet; although it had reverted to its American name before the end of the war. The Martlet entered service towards the end of 1940 and was followed by the first of the successful British naval fighters, the Sea Hurricanes and Seafires, which were navalised versions of RAF aircraft. Although excellent fighters these had certain shortcomings when it came to deck operations. In spite of this, the Seafire in particular was produced in large quantities and used to good effect. But it was really the American carrier aircraft that were leading the way, with the Wildcat, Hellcat, Corsair and Avenger. These were supplied to the Fleet Air Arm under the lease-lend arrangement and proved to be the better naval aircraft of the war, being powerful and robust enough to stand up to the heavy treatment meted out during operations on board carriers.

The Fleet Air Arm had operated successfully in all theatres of war, but perhaps had its greatest success in the Far East during the advance upon Japan, and had proved that it could provide the air support necessary for the air defence of the fleet, in addition to its reconnaissance and strike roles. At the time of the Japanese surrender the Fleet Air Arm had reached its peak, with a total of 59 carriers, 1300 aircraft operated by 69 first line squadrons and another 2500 aircraft operated by second line units. Although the majority of the aircraft carriers were escort types supplied by the USA, usually cargo ships converted primarily by the introduction of a flat top, they had been very effective within the limitations of their main role of defending convoys of merchant ships bringing supplies across the Atlantic from the USA. All six of the armoured fleet carriers survived the war, along with two of the newly

Swordfish of No. 820 and 821 Squadrons preparing for take-off on HMS *Ark Royal* in 1939. (RAF Museum)

Corsair IVs of No. 1831 Squadron, Seafire XVs of No. 806 Squadron and Firefly Is of No. 837 Squadron preparing for take-off from HMS *Glory* shortly after the end of World War Two.

commissioned light fleet carriers, and it was these eight vessels that were to form the basis of the Fleet Air Arm's peacetime force. The escort carriers supplied by the USA under the lease-lend scheme had to be returned to the USA at the end of hostilities. The majority of the American aircraft had also been supplied under lease-lend, but at the end of the war the Americans had no use for all the surplus aircraft so as a means of expediency large numbers of them were dumped in the sea from carrier decks. During the war the Fleet Air Arm's losses were five of the original pre-war fleet carriers, four escort carriers and some 2000 aircraft. However, by the end of the war the success of the aircraft carrier, with its high level of mobility and strike power, resulted in it replacing the battleship as the Royal Navy's capital ship.

The run-down of the Fleet Air Arm following the Japanese surrender was extremely rapid. Within a year it had been reduced to 14 first line squadrons all equipped with British built aircraft, primarily Seafires and Fireflies but with one squadron of the long delayed Firebrands and another with Mosquito FB Mk VIs for training purposes while awaiting the delivery of the first of the Sea Mosquitos. Although wartime operational aircraft were soon replaced by more modern variants or completely new types, a number of second line aircraft continued in service for several years, especially in the training and communications roles. Aircraft such as the Anson, Expeditor, Oxford and Harvard operated until the mid 1950s, and quite remarkably Tiger Moths and Dominies were not replaced until ten years later.

In 1947 the superb Hawker Sea Fury, the first British aircraft with power folding wings, started to enter service and was subsequently to equip eight first line squadrons. It was, however, on the 3 December 1945 that the first pointer to one of the major changes to take place with the equipment of the Fleet Air Arm, was seen when Lt. Cdr. Eric M. Brown carried out the first carrier trials on HMS *Ocean* with a jet powered aircraft, a de Havilland Vampire. However, it was August 1951 before the Fleet Air Arm commissioned its first front line squadron equipped with Supermarine Attacker jet fighters.

When North Korea invaded South Korea on 25 June 1950 it was HMS *Triumph*, equipped with Seafire FR Mk 47s (No. 800 Squadron) and Firefly FR Mk 1s (No. 827 Squadron) that became Britain's first contribution to the United Nations Force which had been mustered to defend South Korea. Although used to good effect,

the Seafires were not sufficiently robust to stand up to the rigours of operating in the conditions prevailing around Korea and consequently the carriers that subsequently operated in those waters were equipped with Sea Furies and later versions of the Firefly. For the three years of the war a British or Australian aircraft carrier was kept on station in Korean waters. On 9 August 1952 Lt. P. Carmichael of No. 802 Squadron, flying a Sea Fury, succeeded in shooting down a Chinese Mig 15 jet fighter. During the Korean War the Fleet Air Arm flew a total of 23000 operational sorties for the loss of 22 aircrew in operations and eleven in accidents.

The Korean War was one of a series of actions that have occurred since the Second World War, and all have demonstrated the flexibility and effectiveness of the Fleet Air Arm. The first of these was the Malayan Emergency which had started in 1948 with communist terrorists trying to gain control of the country defended by government forces supported by units from several Commonwealth countries. The Emergency was to continue for eight years but it was not until the end of 1952 that the Fleet Air Arm became involved in the conflict, forming its first front line helicopter squadron, No. 848, equipped with Whirlwind HAR Mk 21s, specifically to support British Forces in Malaya. It proved very effective in transporting and supplying the units operating in the Malayan jungle. The success of these operations earned it the prestigious Boyd Trophy for 1953. The squadron remained in Malaya until the end of the Emergency in 1956.

A few months later, in November 1956, the Fleet Air Arm was operating five aircraft carriers in Operation Musketeer, the Anglo-French Intervention at Suez. Three of the carriers, HMS *Eagle,* HMS *Albion* and HMS *Bulwark,* operated a total of six Sea Hawk, five Sea Venom, two Wyvern and one Gannet squadrons supported by three Skyraider AEW flights. The other two carriers, HMS *Theseus* and HMS *Ocean* were used as commando carriers providing helicopters to land and then supply the Marine Commando units. Although the operation proved to be a considerable military success, it was brought to a premature end before all the military goals could be achieved by political pressure from the USA and Russia.

In 1961 General Kassem, the ruler of Iraq, began to threaten its neighbour, the small but very rich Sheik-dom of Kuwait, and at the request of the Sheik of Kuwait the British Government went to its defence,

Tiger Moth T Mk 2 on the flight deck of HMS *Eagle.* (B. J. Lowe)

HMS *Ocean*, off Korea in 1952 with No. 802 Squadron Sea Fury FB Mk 11s and No. 825 Squadron Firefly FR Mk 5s ranged on the deck.

sending HMS *Bulwark* with 42 Commando supported by HMS *Victorious* providing air cover. The show of strength appeared to be enough to discourage Iraq and the threat was soon withdrawn. However, before the end of the following year, 1962, the Fleet Air Arm was in action again when guerrillas from Indonesia infiltrated Brunei, and at the request of the Sultan of Brunei the British Government ordered HMS *Albion*, equipped with Nos. 845 and 846 Squadrons operating Wessex and Whirlwind helicopters and carrying No. 40 Commando, to Kuching. This action was very similar to the Malayan Emergency, and continued until 1966. During this time both squadrons were to be awarded the Boyd Trophy, No. 846 in 1963 and No. 845 in 1964.

During the mid 1960s the Fleet Air Arm was involved in several other actions, which included the putting down of an army mutiny in East Africa early in 1964, in which both HMS *Albion* and HMS *Centaur* took part. Shortly afterwards, in May 1964, trouble broke out between the South Arabian Federation and the Yemen, and the Wessex helicopters of No. 815 Squadron from HMS *Centaur* were used to land and then support No. 45 Marine Commando. The last major task undertaken by the Fleet Air Arm in the 1960s was the Beira patrol, which started in March 1966 following the United Nation's resolution imposing an embargo on oil bound for Rhodesia, which had just unilaterally declared itself independent, an action that had been unacceptable to the rest of the world. A small fleet of Royal Navy frigates, headed by one of the Fleet Carriers, patrolled the Mozambique Channel to maintain a blockade of the oil tankers attempting to reach Beira, the nearest port to Rhodesia.

Construction of a number of the carriers laid down during the war was to continue, and subsequently five of the wartime fleet carriers had been replaced during the mid 1950s by two new fleet carriers, HMS *Eagle* and HMS *Ark Royal*. One of the wartime fleet carriers, HMS *Victorious*, was extensively modernised during the 1950s and was to remain in service until the late 1960s. In addition two small fleet carriers, HMS *Albion* and HMS *Bulwark*, were commissioned during 1954, followed by HMS *Hermes* in 1959.

During the 1950s Britain was to develop three major advances in carrier design that would significantly improve safety, in addition to allowing the use of the new heavy, high performance aircraft that were currently being developed. The first of these improvements was the angled deck which had been proposed by Capt. D. R. F. Cambell DSC, RN, and Mr. L. Boddington of the Royal Aircraft Establishment (RAE). The idea was that by angling the landing direction by just a few degrees it was possible to separate the flying section of the deck from the parking area, so that when an aircraft failed to pick up an arrester wire it was possible for it to overshoot and go round again for a further attempt rather than run into the crash barrier or the aircraft in the forward park. Initially the angled deck layout was painted on the deck of HMS *Triumph* for touch and go landings, which confirmed the basic idea. The concept was adopted by both the Royal and US Navies and it was in fact the USS *Antietam* that was the first carrier to be fitted with a fully angled deck. HMS *Centaur* was the

HMS *Bulwark*, at speed, with a Grumman Avenger on the deck. (RAF Museum)

first Royal Navy carrier to be fitted with an angled deck.

In parallel with the development of the angled deck, the RAE was working on a project for a mirror landing aid, which had initially been suggested by Cdr. H. C. N. Goodhart. This equipment was to give the pilot continuous visual information on his position relative to the desired approach path. The initial trials were carried out on board HMS *Illustrious* and HMS *Indomitable* with considerable success, and shortly afterwards the system was adopted by the Royal and US Navies, replacing the deck landing control officer (DLCO), usually referred to as the 'batsman'. The system was subsequently improved by replacing the mirror by a projector and in this form the system is still in use today on conventional carriers.

The third major improvement of carrier equipment came with the steam catapult, which had been designed and developed by Mr. C. C. Mitchell and was to provide a significant increase in accelerating force, which had become progressively more important as the higher performance aircraft entered service during the 1960s and 1970s. Early trials with the steam catapult were carried out aboard HMS *Perseus* during 1950/51. The steam catapult was also adopted as standard equipment by the US Navy.

The Fleet Air Arm had started to take an interest in helicopters towards the end of the Second World War, and early in 1945 a small number of R-4B Hoverflies were put on the strength of No. 771 Squadron for trials and a limited pilot training role. Although the Hoverfly had no operational use, it did provide the Fleet Air Arm with experience of helicopters and a knowledge of the potential roles that could be covered by more advanced types. In September 1946 a Hoverfly was landed on the frigate HMS *Helmsdale* by Lt. Alan Bristow — who was later to become well known as the founder of Bristow Helicopters Ltd. On 1 February 1947 a Hoverfly flown by Lt. K. Reed made a first landing on a battleship, when it landed on HMS *Vanguard*. In May 1947 the Royal Navy formed its first helicopter squadron, No. 705, equipped with Hoverflies, which became responsible for all the Fleet Air Arm's helicopter activities, including training, trials and development work.

The next helicopter on the scene was the Sikorsky S-51, which became known as the Dragonfly and was developed and built under licence by the Westland Aircraft Company. No. 705 Squadron had re-equipped with Dragonfly HR Mk 1s by June 1960 for pilot training, but these had been replaced by the more suitable Hiller HT Mk 1s in the pilot training role in 1953, although Dragonflies were retained for other duties by the squadron. The Dragonfly was also used by the various ships and station flights for air sea rescue duties, replacing the wartime Sea Otters. The Fleet Air Arm was eager to obtain a helicopter capable of meeting its requirements for the anti-submarine role. This entailed carrying both the dunking sonar detection equipment and the anti-submarine weapons. The Whirlwind lacked the power to carry both the sonar equipment and weapons and its successor, the WS-58 Wessex, could just about carry both, but fully loaded its endurance was totally inadequate. Both types had proved very useful in service, with variants covering anti-submarine, air sea rescue and utility roles. It was not until the arrival of the WS-61 Sea King in 1969 that

the Fleet Air Arm had a helicopter which would satisfactorily carry out both tasks. The Whirlwind was phased out of service in 1977 and the Wessex in 1988. In the early 1950s, the Fleet Air Arm was considering using a development of the twin rotor Bristol type 173 helicopter for the anti-submarine role, but development problems soon caused them to lose interest in the project. The Sea King is currently the Navy's prime anti-submarine and cargo/troop carrying helicopter and is likely to remain so well into the 1990s when it will be replaced by the EH101, the prototype of which was rolled out of Westland's Yeovil factory on 7 April 1987. The small Westland Wasp, which has been operating in the anti-submarine role from the Royal Navy's frigates since the mid 1960s, has now been completely replaced by the Westland/Aerospatiale Lynx, one of the three Anglo/French collaborative helicopter projects. The only other helicopter currently in use by the Fleet Air Arm is the Aerospatiale/Westland Gazelle, which is used for pilot training by No. 705 Squadron.

In 1946 the Admiralty decided to form the Air Branch of the Royal Naval Volunteer Reserve (RNVR) and No. 1831 Squadron, under the command of Lt. Cdr. N. G. Mitchell DSC, was the first to form at RNAS Stretton on 1 June 1947, with six Seafire FR Mk XVIIs and two Harvards. This was followed later in the year by two more fighter squadrons, Nos. 1832 and 1833 at RNAS Culham and Bramcote respectively, and No. 1830 Squadron, a combined fighter and anti-submarine squadron, which had reformed at RNAS Abbotsinch with three Seafire Mk XVIIs and three Firefly Mk 1s. In 1949, the squadrons held their annual two weeks training aboard carriers with Nos. 1830, 1831 and 1833 using HMS *Illustrious* and No. 1832 using HMS *Implacable*. In 1951 there was the start of a

HMS *Victorious*, HMS *Ark Royal* and HMS *Hermes* in the Mediterranean during the summer of 1960. (Lt. Cdr. M. S. Lay)

HMS *Eagle* in March 1970 with Sea Vixen FAW Mk 2s of No. 899 Squadron, Buccaneer S Mk 2s of No. 800 Squadron, a Gannet COD Mk 4 of No 849 Squadron and a Wessex of the Ship's Flight.
(Rolls Royce)

major expansion of the RNVR to meet the Admiralty's demand for more anti-submarine squadrons, and progressively each existing squadron had an anti-submarine squadron reformed alongside it. In 1952 these two-squadron groups were each given divisional status with the Scottish Air Division at Abbotsinch, the Northern Air Division at Stretton, the Midland Air Division at Bramcote, the Southern Air Division at Culham and the Channel Air Division at Ford.

By May 1954 all of the RNVR fighter squadrons had replaced their Seafires with Sea Furies. Shortly afterwards, in May 1955, No. 1831 Squadron at Stretton re-equipped with Attackers, becoming the first RNVR jet fighter squadron. The rest of the RNVR fighter squadrons had re-equipped with Attackers before the end of 1955, and the following year No. 1832 at RAF Benson had re-equipped with Sea Hawk F Mk 1s. Re-equipment of the RNVR anti-submarine squadrons started in 1955 when No. 1830 Squadron of the Scottish Air Division re-equipped with Avenger AS Mk 5s, and was followed shortly afterwards by Nos. 1841 and 1844 Squadrons, while No. 1840 Squadron of the Channel Air Division at Ford re-equipped with Gannet AS Mk 1s. The plan was that by the end of 1957 all the fighter squadrons would be re-equipped with Sea Hawks and

all the anti-submarine squadrons with Gannets. However, this was not to be, for towards the end of 1956 the government decided to disband the Air Branch of the RNVR as an economy measure and all the squadrons officially disbanded with due ceremony on 10 March 1957.

In 1959 two of the Royal Navy's carriers, HMS *Albion* and HMS *Bulwark* were converted to commando carriers and, as they were intended for helicoper operations only, the equipment for fixed wing operations, including arrester wires, catapult and mirror

Buccaneer S Mk 2 of No. 800 Squadron landing on HMS *Eagle* in April 1970. (Rolls Royce)

landing aid, was deleted. The following year the prospects for the Fleet Air Arm looked good when the government announced the Royal Navy's new carrier CVA-01. The mood in the Fleet Air Arm was to change to despair some ten years later when the government of the day cancelled CVA-01, causing the resignation of the First Sea Lord and the Navy Minister. As the Navy's existing carriers were coming to the end of their service, it seemed inevitable that operational fixed wing flying in the Fleet Air Arm would come to an end, restricting it to helicopter operations with air cover for the fleet having to be provided by the land based aircraft of the RAF. HMS *Victorious* was the first of the carriers to be taken out of service when a small fire on board during her final refit in 1967 led to a decision to cancel the refit. The next was HMS *Hermes* which was converted into a commando carrier in 1971, leaving just the two Fleet carriers, HMS *Ark Royal* and HMS *Eagle* to carry on until the mid-1970s. However, HMS *Eagle* was taken out of service in 1972. HMS *Ark Royal* was re-commissioned on 24 February 1970, following a major refit at Devonport Dockyard, and with her carrier air group comprising Phantoms (No. 892 Squadron), Buccaneers (No. 809 Squadron), Gannet AEWs (No. 849B Flight) and Sea King ASWs (No. 824 Squadron) was the most powerful carrier ever operated by the Fleet Air Arm. Before she was withdrawn from service in December 1978, HMS *Ark Royal* had made several trips to the Mediterranean, Caribbean and the USA.

A major step forward in the defence of the fleet from air attack came about in 1952 when No. 849 Squadron reformed with Skyraider AEW Mk 1s. This squadron was divided into flights so that one could be attached to

each of the carriers. Rather belatedly the Wyvern strike aircraft started to replace the Firebrand in 1953, and the same year the Sea Hawk entered service and was subsequently to replace the Attacker. Delays with the development of the anti-submarine Gannet resulted in the Fleet Air Arm obtaining 100 Avengers in 1953 from the USA until the Gannet became available. The two-seat de Havilland Sea Venom all-weather fighter started to enter service in March 1954 and the Gannet made its first appearance in January 1955. The Gannet's career in the anti-submarine role was cut short by the success of the helicopter in this role, and by 1960 it had been almost entirely replaced by the Westland Whirlwind.

After a lengthy period of development, the Supermarine Scimitar started to replace the Sea Hawk in 1958, and for the first time the Fleet Air Arm was to receive a really effective jet fighter: well equipped and with a good performance. The Sea Vixen started to replace the Sea Venom in the all-weather fighter role and when the Scimitar was phased out of service, the Sea Vixen took over the total air defence role on all the carriers, except for HMS *Ark Royal*. Shortly after the purchase of the McDonnell Phantom, the decision was taken to cancel CVA-01 and, as only HMS *Ark Royal* and HMS *Eagle* were planned to operate to the mid 1970s, both carriers were to be modified to take the Phantom. In the event a change of plan resulted in HMS *Eagle* being taken out of service earlier than planned, and consequently only HMS *Ark Royal* was modified. The sole Fleet Air Arm Squadron, No. 892, was to operate as part of HMS *Ark Royal*'s air group until the ship was taken out of service in December 1978, when the surviving Phantoms and Buccaneers were handed over to the RAF.

Following the cancellation of CVA-01, the Admiralty started to re-assess the requirements of the Fleet Air Arm as a purely helicopter equipped force. This work culminated during the late 1960s in a new class of ship

Phantom FG Mk 1 of No. 892 Squadron on the catapult of HMS *Ark Royal* with Buccaneer S Mk 2s of No. 809 Squadron ranged under the Island, in 1975. (Lt. Cdr. M. S. Lay)

One of the two Sea Harriers kept at readiness for air defence duties on board the ill fated *Atlantic Conveyor*, seen taking-off while the ship was anchored off Ascension during its voyage to the South Atlantic as part of the Falklands Task Force. (F.P.U.)

to operate helicopters and known at the time as a through-deck or anti-submarine cruiser, although it has since been re-designated a carrier vessel submarine attack (CVSA), although usually referred to as an ASW carrier. Three through-deck cruisers were ordered and the first, HMS *Invincible*, was launched in 1977 followed later by HMS *Illustrious* and HMS *Ark Royal*. The Admiralty, seeing a chance to get back into fixed wing flying, proposed the development of the RAF's Harrier to operate from the ASW carrier as part of an air group, along with anti-submarine Sea Kings. The lack of deck space was likely to prevent the optimum STOL operation of the Harrier on board ship, but this problem was resolved by a Royal Navy engineering officer, Lt. Cdr. D. R. Taylor, who prepared a thesis at Southampton University on the advantages gained by launching a STOL type aircraft on an upward trajectory by means of a curved ramp. Trials proved the benefit of the 'ski-jump' launch and consequently ramps were introduced on the forward end of the deck of the ASW carriers and an additional one was also fitted to the front of HMS *Hermes*, which was operating as an ASW (helicopter) carrier. The Sea Harrier was ordered in 1975 and entered squadron service in 1980.

Early on the morning of 2 April 1982, a large Argentinian invasion force landed on the Falkland Islands and, as the small detachment of Royal Marines based on the Islands was totally outnumbered, the Argentinians gained control before the end of the day. The British Government's response to this outrage was virtually instantaneous and almost immediately a task force was being prepared to sail to the South Atlantic to recover the Falkland Islands. The operation, given the name "Corporate", was put under the overall command of Admiral Sir John Fieldhouse GCB, and was to involve 33 Royal Navy surface ships, six submarines, 24 Royal Fleet Auxiliaries, five trawlers and 45 merchant ships. By the end of April some 70 ships of the task force had put to sea. Helicopter landing platforms were fitted to a number of the larger merchant ships, primarily to improve communications but also as alternative platforms for operational aircraft, a few being capable of use by Sea Harriers in addition to the helicopters. The Fleet Air Arm provided 28 Sea Harriers and 150 helicopters ranging from the obsolescent Wasp and Wessex to the latest versions of the Sea King and Lynx. Fortunately, HMS *Hermes* and the first of the new ASW carriers, HMS *Invincible*, were operational and became major elements in the task force.

The task force succeeded in recovering the Falkland Islands, with the Argentinian commander signing the surrender documents on 14 June 1982. Although all the constituents of the task force must share in the success of Operation Corporate, it is generally accepted that the operation could not have succeeded without the magnificent Sea Harriers and the two carriers from which they operated. The Sea Harriers and their pilots out-performed all the enemy aircraft they encountered and of the six Sea Harriers lost, four were lost in accidents and two were shot down by ground fire and this compares with 24 Argentinian aircraft actually shot down by Sea Harriers. Unfortunately, a number of British ships were sunk or badly damaged, and this to some extent could be put down to a lack of Airborne Early Warning cover that had been provided by the Gannet AEW Mk 3s of No. 849 Squadron until the demise of the fleet carrier HMS *Ark Royal* in 1978. This deficiency has since been resolved by the development of the Sea King AEW Mk 2 equipped with Searchwater radar, which is now operating from the current ASW carriers.

Since the Falklands War, HMS *Hermes* has been sold to India, becoming the INS Viraat, now that all three of the ASW carriers, HMS *Invincible,* HMS *Illustrious* and HMS *Ark Royal,* have been commissioned. The future of helicopters in the Fleet Air Arm is obviously secure and, following the success of the Sea Harriers in the South Atlantic, it is hoped that the politicians are now convinced that a fixed wing air defence and strike element is essential for fleet operations.

Sea Harrier FRS Mk 1s and Harrier GR Mk 3s wrapped up to protect them from the weather on the deck of *Atlantic Conveyor* for delivery to the South Atlantic. The Sea Harrier on the platform on the bow of the ship was kept operational to provide air defence when in range of the Argentine aircraft. (B.Ae.)

CHAPTER ONE

Supermarine Seafire (Griffon Engine)

After trials with the Sea Hurricane had proved that high performance aircraft could be operated successfully from a carrier deck, the Royal Navy started to consider the feasibility of adapting Spitfires for carrier operations.

Initially, a Spitfire Mk VB AB205 was fitted with an arrester hook under the fuselage and used for trials aboard HMS *Illlustrious* early in 1942. Following the success of these trials, authority was given to proceed with the Sea Spitfire (later contracted to Seafire). Over 150 Mk VB and VCs were initially converted into Seafire Mk Is for the Royal Navy at Air Service Training and Supermarine, by the introduction of an arrester hook, slinging points and some reinforcing of the fuselage. The Seafire Mk I was further developed into the Mk II by the introduction of catapult spools and additional strengthening of the fuselage and undercarriage. However, because of the lack of folding wings on the early Seafires, their operational use on carriers was rather limited. It was not until the introduction of the double folding wing on the Seafire Mk III that full normal carrier operations could be carried out with the type, allowing for easier deck handling and the use of the lift and storage on the hangar deck. The Mk III was also the first of the Seafires to be built from new, rather than by conversion of existing RAF Spitfires. It proved to be highly successful and was built in large quantities, a total of 1263 having been completed by the end of the war. The type served with 25 of the Fleet Air Arm's front line fighter squadrons in addition to an even larger number of second line squadrons.

Seafires were involved in several of the important wartime operations in Europe, generally providing top cover for the sea-borne landings during the Allied invasions of North Africa (Operation Torch), Italy (Operation Avalanche) and the South of France (Operation Dragoon). By VJ-Day there were eight Seafire squadrons in the Far East, all operating Mk IIIs from six Royal Navy aircraft carriers. These squadrons were used to provide fighter cover for air attacks on various targets during the Allied advance on Japan.

Impressed with its operational success, the Royal Navy considered that a version of the Seafire with an improved performance would be ideal for the Far East. To meet this requirement, Supermarine proposed a development of the Seafire Mk III to be powered by a Rolls Royce Griffon engine with a two-speed, single-stage blower. To cover this project, identified as the Type 377 by Supermarine, Specification N.4/43 was issued, followed up in March 1943 by an order for six prototypes. This version of the Seafire, the Mk XV, was basically a Seafire Mk III airframe fitted with a retractable tailwheel and powered by the 1750hp Griffon Mk VI engine driving a Rotol four-bladed propeller. In July 1943, the first production order was placed with Cunliffe-Owen for 150 Seafire Mk XVs, followed in February 1944 by an order for another 140 to be built by the Westland Aircraft Company, who also received a contract the following month for a further 500 aircraft. However, with the end of the war the programme was cut back and in the event only five prototypes and 384 production aircraft were built. The first 50 of these retained the A-frame arrester hook, but later aircraft had a new 'sting' type arrester hook fitted to the rear of the fuselage under the rudder.

The prototype Mk XV, NS487, was flown for the first time in February 1944 and within a month it joined No. 778 Squadron, the Service Trials Unit at RNAS Crail,

Seafire F Mk XV PR370/515/LP of No. 773 Squadron, Lee-on-Solent, at Gibraltar in March 1950. (RAF Museum)

Seafire FR Mk 47 of No. 804 Squadron on the lift aboard HMS *Ocean* in the Mediterranean during 1948.

built by Cunliffe-Owen, and this was followed by 25 production aircraft built at the Vickers-Armstrongs South Marston factory and delivered between November 1945 and May 1947. The production aircraft differed from the prototype only with the installation of a Griffon 88 in place of the Griffon 87. Like its predecessor the Mk 45, the Mk 46 was used by various second line squadrons, the last to use the type being No. 738, the Naval Air Fighter School, which phased its last Mk 46s out of service in August 1950. When No. 1832 RNVR Squadron reformed at RNAS Culham in August 1947, it was initially equipped with a mixture of Seafire L Mk IIIs and F Mk 46s. The F Mk 46s were to remain in service alongside the L Mk IIIs and subsequently the Seafire F Mk XVIIs until January 1950.

The final version of the Seafire to be built for the Royal Navy was the FR Mk 47, and this differed from the F Mk 45 and F Mk 46 primarily by the introduction of folding wings. On the early production aircraft the wings were folded manually, but later hydraulic power folding was introduced. The wing hinge line was outboard of the wheel wells, and the guns, mounted further outboard than on previous Seafires, obviated the necessity for the wingtips to fold downwards, as did the earlier variants, to keep within the height limitations imposed by the hangar decks on carriers. Extra internal fuel tanks were introduced in the leading edge of the wing and in the rear fuselage, and fuel capacity could be further increased by the installation of a 90 gallon drop tank under the fuselage and two 23 gallon combat tanks under the wings.

An order was placed for 150 Seafire Mk 47s during 1946, but this was soon reduced to 90, all of which were built at Vickers-Armstrongs South Marston factory between April 1946 and January 1949. The first production aircraft, PS944, was retained for development purposes. Most of the Seafire Mk 47s were built as fighter reconnaissance versions, identified as FR Mk 47s, and fitted with provision in the rear fuselage for two electrically heated cameras, one vertical and one oblique. Deliveries of the Seafire Mk 47 began with No. 778 Squadron, who undertook service trials between December 1946 and March 1947. Trials were also carried out by No. 787 Squadron, the Naval Air Fighting Development Squadron of the Central Fighter Establishment at RAF Tangmere who operated Mk 47s

from May 1947 until September 1949. The only other second line squadron to use this version was No. 759, which operated as part of the Naval Air Fighter School at RNAS Culdrose and used the Mk 47s along with Seafire F Mk XVIIs and Firebrand TF Mk Vs from September 1952 until November 1953.

However, unlike the previous two marks of Seafire, the FR Mk 47 was operated by two front line squadrons and also proved to be the only version of the Griffon powered Seafires to fire its guns in anger. No. 804 Squadron, commanded by Lt. Cdr. S. F. Shotton DSC, was the first to re-equip with the Seafire FR Mk 47 when it replaced the F Mk XVs with 13 FR Mk 47s in January 1948. The Squadron spent the next eight months working up ashore before embarking on HMS *Ocean* on 24 August 1948 for service in the Mediterranean. The Seafire FR Mk 47s were subsequently replaced by Sea Furies while No. 804 Squadron was based at RNAS Hal Far. The other front line squadron to operate the Seafire FR Mk 47 was No. 800 which received its aircraft in April 1949, replacing its existing Seafire F Mk XVIIs.

The squadron almost immediately embarked on the light fleet carrier HMS *Triumph* which sailed by way of the Mediterranean for the Far East, where the squadron carried out strikes against the Malayan terrorists during the second half of 1949 and early 1950. Then, when North Korea attacked South Korea in June 1950, the British Eastern Fleet, headed by HMS *Triumph* was diverted to Korean waters. During her tour of duty off Korea, HMS *Triumph* completed three patrols and although there was a general shortage of both spare parts and replacement aircraft, the Seafires of No. 800 Squadron were very effective, flying some 360 sorties including 115 strafing, bombing and rocket attacks on communications targets on the Korean mainland. Fortunately the squadron suffered very few casualties, although in August the Squadron Commander, Lt. Cdr. I. M. MacLachlan, was killed in an accident on the flight deck and was replaced by Lt. T. D. Handley. Despite intensive work by the maintenance crews, the squadron was eventually reduced to a single operational Seafire by the end of September 1950. Fortunately HMS *Triumph* was relieved by HMS *Theseus* the following month and returned to the UK where No. 800 Squadron disbanded on 10 November 1950, ending the Seafire's first line service.

The only RNVR squadron to use the Seafire Mk 47 was No. 1833 Squadron at RNAS Bramcote, which replaced its Seafire F Mk XVIIs with ten Seafire FR Mk 47s in June 1952. These in turn were replaced by Sea Furies in May 1954, just two months before the last of the Seafire F Mk XVIIs were phased out of service by No. 759 Squadron. By this time the Seafire had completed twelve years of distinguished service as a Fleet Air Arm fighter.

In 1942 Supermarine started work on a laminar flow wing for use on a projected Spitfire replacement, Specification F.1/43 subsequently being issued to cover the project. Initially plans were put in hand to fit this new laminar flow wing to the Spitfire F Mk 21, and this new project was named the Spiteful. At the same time, Supermarine proposed a development of the Seafire F Mk XV fitted with the new laminar flow wing and powered by a Griffon 61 engine. No action was taken, however, until 1945 when Specification N.5/45 was issued for a single-seat fighter to replace the Seafire. In

The first Seafang F Mk 31, VG471. (RAF Museum)

April of that year two prototypes of the Supermarine project, named the Seafang, were ordered and this was followed shortly afterwards by an order for 150 production aircraft. As part of the Seafang development programme, the sixth production Spiteful F Mk 14, RB520, was fitted with a sting type arrester hook and used by Supermarine for trials. Of the two prototypes ordered one, VB893, was cancelled but the other, VB895, was completed and designated Seafang Mk 32. This aircraft had hydraulically operated folding wings and was powered by a Griffon 89 driving a contra-rotating propeller. The first production Seafang was completed late in 1945 and was designated the F Mk 31 as it did not have folding wings and was fitted with a five bladed propeller.

The armament of the Seafang was primarily four 20mm cannons mounted in the wings with provision to carry two 1000lb bombs or six 300lb rocket projectiles under the wings. The aircraft was fitted with a sting type arrester hook, slinging points and fittings for rocket assisted take-off. The wing fold lines were well outboard so that little more than the wing tips folded upwards.

Carrier trials were carried out by Lt. Cdr. M. J. Lithgow using VB895, who made eight landings on board HMS *Illustrious* on 21 May 1947. As a result the pilot concluded that the Seafang was more suitable for carrier operation than any of the Seafire variants! However, by this time the Royal Navy was losing interest in the Seafang as it became apparent that jet fighters would supersede the piston engined types in the not too distant future and the Seafire Mk 47s and Sea Fury FB Mk 11s would provide adequate cover in the interim period. As a consequence the production order was reduced to 16 aircraft, and in the event only nine were actually completed.

Supermarine's jet fighter project, the Type 392 which was later to become the Fleet Air Arm's Attacker, was fitted with what were basically Seafang wings, and consequently it was possible to use some of the Seafangs in the Type 392 flight development programme.

1 Seafire FR Mk 47

Production

F Mk XV NS487, NS490, PK240, PK243, PK245 (Prototypes).
 PR338 – PR379, PR391 – PR436, PR449 – PR479, PR492 – PR516, SR446 – SR493, SR516 – SR547, SR568 – SR611, SR630 – SR645, SW781 – SW828, SW844 – SW875, SW876 – SW879, SW896 – SW921.
F Mk XVII NS493 (Prototype).
 SP323 – SP327, SP341 – SP355, SW986 – SW993, SX111 - SX139, SX152 – SX201, SX220 – SX256, SX271 – SX316, SX332 – SX370, SX386 – SX389.
F Mk 45 LA428 – LA457, LA480 – LA499.
F Mk 46 LA541 – LA564.
F Mk 47 PS944 – PS957, VP427 – VP465, VP471 – VP495, VR961 – VR972.
Seafang F Mk 31 VG471 – VG480.
 F Mk 32 VB895 (Prototype).
 VG481, VG482, VG486, VG488 – VG490.

CHAPTER TWO

Blackburn Firebrand

The development programme that was ultimately to lead to the Firebrand aircraft started in March 1939 when the Air Ministry issued Specifications N8/39 and N9/39 for two-seater, short range, ship-borne interceptors; the N9/39 specification differing from the N8/39 in its requirement for a four-gun, power-operated turret. However, the concept of fighters equipped with gun turrets soon fell from favour. The

Firebrand TF Mk II, DK383/OC, of No. 708 Squadron, RNAS Lee-on-Solent. (B.Ae.)

correctness of this decision was confirmed by the dismal failure of the Boulton Paul Defiant and the Blackburn Roc. As a result the N9/39 specification was rewritten the following year and issued as N5/40, which produced the highly successful Fairey Firefly two-seater fleet fighter.

In the meantime the Specification N8/39 requirement was changed to a single-seat fighter and a new specification, N11/40, was issued to cover it. Both the Blackburn and Hawker Aircraft Companies submitted

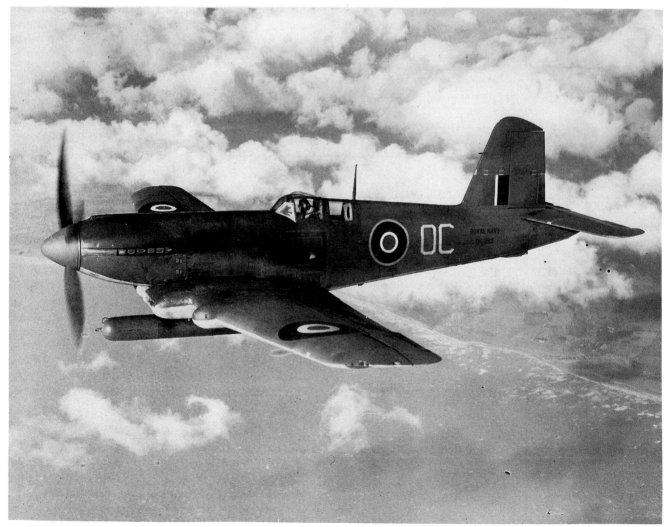

Late production Firebrand TF Mk 4, EK726, on test flight from Brough. (B.Ae.)

projects to meet the specification but, presumably because of Hawker's very heavy commitments at this stage of the war, little consideration appears to have been given to their proposals and in July 1940 an order was placed for three prototypes of the Blackburn B-37 project. Blackburn's chief designer, G. E. Petty, and his team immediately started work on the project. By September a full scale mock-up had been built and on 11 July 1941 the B-37 was given the name Firebrand. The first prototype, DD804, which was in essence the aerodynamic prototype as it carried no armament, made its maiden flight from RAF Leconfield on 27 February 1942 piloted by Flt. Lt. A. Thompson.

The Firebrand was a single seat, all-metal, low-wing monoplane, very large for an interceptor and fitted with a 2305hp Napier Sabre III twenty-four cylinder H-type engine. By housing the engine coolant radiators in extensions to the centre-section leading edges, it was possible to provide a very neat, streamlined engine cowling. The airframe was of conventional stressed skin construction, except for the forward fuselage, which was a tubular-steel structure covered with detachable aluminium alloy panels.

Manufacturer's trials at Brough resulted in modifications being carried out to the tailplane and elevators before it was delivered in June 1942 to the A&AEE, Boscombe Down, for performance trials. Whilst DD804 was at the A&AEE, the second prototype, DD810, flew for the first time on 15 July 1942. This aircraft was fully armed with four 20mm Hispano cannons mounted in the outer wing, along with provision for bomb racks to carry up to four 500lb bombs. DD810 was delivered to the Royal Navy on 11 October 1942 and, whilst operating from RNAS Machrihanish, carried out carrier trials on board HMS *Illustrious* during February 1943. The third and last prototype, DD815, was completed shortly after DD810 and was delivered to the A&AEE for armament trials on 15 September 1942.

On completion of the contractor's trials the first prototype, DD804, was delivered to the test flight centre of D. Napier and Son Ltd. at Luton on 25 May 1943, where a completely revised engine installation was developed. However, at an early stage the production of these new units was cancelled when the Ministry of Aircraft Production (MAP) decided that

future production of Sabre engines was to be allocated to the Hawker Typhoon. By this time, however, the writing was on the wall for the future of the Firebrand as a ship-borne interceptor, the successful introduction of the Supermarine Seafire having shown that the Firebrand was already completely outclassed as a fighter.

The MAP, apparently not wishing to waste all the development that had already been put into the Firebrand, asked Blackburn's to redesign the aircraft to meet a requirement for a torpedo-fighter. The large size of the Firebrand made its conversion to the torpedo-fighter role relatively simple, but even so it necessitated widening the centre-section by 1ft 3.5in to provide clearance between the wheel wells for an 18in torpedo, and this increased the wingspan to 51ft 3.5in. To expedite the development of the Firebrand torpedo-fighter, work was put in hand to convert the second prototype, DD810, into the prototype of the Firebrand TF Mk II. This aircraft, re-serialled NV636, flew for the first time at Brough on 31 March 1943.

However, production of the Firebrand F Mk I was allowed to continue, eleven F Mk Is, DK363 to DK373, being produced before the line changed over to the Firebrand TF Mk II. These aircraft, already obsolete, were used for experimental purposes. DK363 was used by Napier for power plant development, DK364 was retained by Blackburn's for about a year until delivered to the RAE at Farnborough on 29 February 1944. DK368 was also retained by the manufacturers at Brough where DK372 and DK373 were also retained as incomplete airframes until used as TF Mk III development aircraft. DK370 was delivered to RNAS Lee-on-Solent for evaluation by the Royal Navy, and the remaining five aircraft of the first batch were all delivered to the A&AEE at Boscombe Down.

The decision that the deliveries of Sabre engines were to be concentrated on meeting Hawker's requirements for the Typhoon was the main cause of the premature end of the production of the Firebrand TF MK II after only twelve aircraft, DK374 to DK385, had been completed, and again these aircraft were only used for trials. In September 1944 the Firebrand Tactical Trials Unit was formed at Lee-on-Solent under the command of Lt. K. Lee-White, initially as No. 764B Squadron, and from 2 October 1944 as No. 708 Squadron. By August 1945, when the squadron replaced its Firebrand TF Mk IIs with the later TF Mk IIIs, they had operated eight of the twelve production TF Mk IIs. In addition to the Firebrands, 708 Squadron also operated several Spitfires which at times were used to develop Firebrand combat tactics and at the same time confirm the Firebrand's general inferiority to the Seafire. During all the time with the squadron the Firebrand proved to be very troublesome, particularly with regard to the Sabre engine which always seemed rather loathe to start, and the consequent lack of serviceable aircraft had a disruptive effect on the progress of the trials. On 6 May 1945 two of the squadron's aircraft, flown by Lt. K. Lee-White and Lt. P. G. Lawrence, carried out deck landing trials aboard HMS *Glory*. These were brought to a premature conclusion when the port undercarriage leg on one of the aircraft collapsed on landing. The trials recommenced later in the month on board the trials carrier HMS *Pretoria Castle* and were completed without further incident.

Following the decision to divert the supply of Sabre engines to Hawkers, an alternative engine was sought and eventually Specification S8/43 was issued on 3 October 1943 for a new version of the Firebrand to be powered by the 2400hp Bristol Centaurus VII, a two-row, eighteen-cylinder, sleeve-valve, radial engine driving a Rotol 13ft 6in diameter, four bladed propeller. During the design of this new engine installation the opportunity was taken to introduce spring loaded tabs on all the control surfaces, a rudder of larger chord and several other minor changes. This project was considered to be sufficiently different to justify the allocation of a new type number, the B-45, and to be re-designated the Firebrand TF Mk III, and two of the unfinished production F Mk Is were completed as the prototypes of the TF Mk III. The first of these, DK372, flew for the first time at Brough on 21 December 1943 and, after a period at the A&AEE, went to Rotol Limited at Staverton in March 1945 to be used for propeller development. The second prototype, DK373, also went to the A&AEE before going to the RAE at Farnborough on 2 June 1945 for radio altimeter trials.

The first production Firebrand TF Mk III, DK386, made its initial flight at Brough with Sqn. Ldr. J. R. Tobin at the controls in November 1944, and all 27 production aircraft had been completed by May 1945. Five of the TF Mk IIIs were powered by the Bristol Centaurus VII engine and the remainder with the improved Centaurus IX mounted on special anti-vibration mountings. None of the Firebrand TF Mk IIIs was to enter operational service, all being used for trial purposes. The last eight were delivered to the Royal

Navy at RNAS Anthorn, from where they were ferried out to the various units. These included No. 700 Squadron, the maintenance test pilots' training squadron commanded by Lt. Cdr. L. R. E. Castlemaine at RNAS Worthy Down, No. 703 Squadron, and the Naval Air Sea Welfare Development Unit (NASWDU), commanded by Lt. Cdr. J. H. Dundas DSC, at RAF Thorney Island. In August 1945 they also replaced the Firebrand TF MK IIs operated by No. 708 Squadron at RNAS Gosport.

The next version, the Firebrand TF Mk IV identified as the type B-46, differed by the inclusion of a dive-bombing capability, which required the fitting of retractable spoilers in the upper and lower wing surfaces and racks under each wing that could carry up to 2000lb of bombs or alternatively sixteen 60lb rocket projectiles. To extend the range, two 45-gallon drop tanks could be carried under the wings or a single 100 gallon drop tank under the fuselage. To improve the handling at low speeds, a larger horn balanced rudder was introduced in addition to a large fin. The fin was offset three degrees to port to reduce the swing on take-off.

A total of 103 Firebrand TF Mk IVs, serial numbers EK601-EK638, EK653-EK694 and EK719-EK741, were built between May 1945 and January 1946, the TF Mk IV thus proving to be the main production variant. Although the majority of the aircraft were delivered direct from the factory to the Fleet Air Arm, a number were retained for trial purposes.

The Firebrand made its public debut at the ATA Farewell Display at White Waltham on 29 September 1945, and this was followed by a display at Heston on 2 October 1945 by two Firebrand TF Mk IVs, EK602 and EK665.

Firebrand TF Mk 4s, EK617 and EK601. (RAF Museum)

Firebrand TF Mk 5s, EK629/102/A, EK734/110/A and EK609/100/A (all conversions from TF Mk 4s), of No. 813 Squadron in November 1952.

The first operational Firebrand squadron, No. 813, was formed at RNAS Ford on 1 September 1945 under the command of Lt. Cdr. K. Lee-White, who has previously commanded the Firebrand Tactical Trials Unit. Equipped with 15 Firebrand TF Mk IVs, the squadron immediately started on a training programme to bring it up to operational status. The squadron's Firebrands made their first public appearance when they were given the honour of leading the Victory Flypast over London on 8 June 1946, and they also took part in the subsequent air display at Eastleigh. At this time EK746 was part of the static exhibition in Green Park, London, and later at the British Aircraft Display at Farnborough on 28 June. The squadron disbanded on 30 September 1946, never having operated on board a carrier.

Development of the Firebrand continued, and by the end of 1945 the TF Mk IV had been superseded on the production line by the TF Mk V, which differed from the Mk IV in having horn-balanced elevators and longer-span aileron tabs. A total of 120 TF Mk Vs were ordered, but with the cessation of hostilities only 67 production aircraft were completed. However, in addition at least 60 TF Mk IVs were converted to TF Mk V standard by the Blackburn Aircraft Company at Brough. The final version of the Firebrand was the TF Mk VA, which was basically a Mk V fitted with hydraulically-boosted ailerons. EK769 was used as the prototype TF Mk VA, and it is believed that only a further six were converted to Mk VA standard. One aircraft was fitted with power assisted elevators, but even though these proved to be satisfactory the modification failed to achieve production status.

Firebrand TF Mk V EK850 was entered in the High Speed Handicap Race at Lympne on 31 August 1947 and was flown into second place at an average speed of 310.69 mph by Group Captain C. J. P. Flood. Two years later, Blackburn's chief test pilot, P. G. Lawrence, flying Firebrand TF Mk V EK621 won the Air League Challenge Cup at Elmdon on 30 June 1949. However, in the same event on 22 July of the following year, P. G. Lawrence flying EK644 lost his title when he was unplaced in the race.

The Firebrand returned to operational service on 1 May 1947 when No. 813 Squadron reformed with twelve Firebrand TF Mk Vs at RNAS Culdrose. The squadron was aboard HMS *Illustrious* in September 1947 for two days of deck landing practice. The Firebrand, however, continued to be most troublesome and for the remainder of 1947 and the whole of 1948 the squadron was only able to undertake a further two periods aboard HMS *Implacable* totalling some nine weeks. In May 1948 the squadron took part in Exercise "Dawn" operating from RNAS Arbroath, and on 10 June 1948 was part of the King's Birthday Review Flypast at RNAS Gosport, followed two days later by the Air Display at RNAS Bramcote. In June 1948, with the official move from Roman to Arabic numerals for aircraft mark numbers, the Firebrand TF Mk V was re-designated the TF Mk 5, although this change does not appear to have been applied to the earlier marks.

November 1948 saw the formation of the 1st Carrier Air Group (CAG) with No. 813 Squadron as the strike element and No. 801 Squadron, equipped with Sea Hornet F Mk 20s, as the fighter element. No. 813 Squadron, as part of the 1st CAG, spent a total of six months during 1949 aboard HMS *Implacable* for cruises in Home, Mediterranean and Norwegian waters. Except for a change to HMS *Indomitable,* the squadron's activities continued in a similar vein until February 1953, when it moved to RNAS Ford to prepare for the arrival of the much delayed Westland Wyverns. Re-equipment commenced in May 1953 and was completed in August 1953, ending six years of Firebrand first line service.

Only one other first line squadron operated the Firebrand, and that was No. 827 which, after a period as part of the 13th CAG in the Far East equipped with Fairey Fireflies, returned to the UK in November 1950 intending to re-equip with the Westland Wyvern. The Wyvern, however, was still involved in trials and it was decided to re-equip with Firebrand TF Mk 5s. Consequently on 13 December 1950, under the command of Lt. Cdr. R. Henderson, No. 827 Squadron reformed at RNAS Ford as only the second Firebrand squadron. After an initial working-up period at Ford, the squadron flew to Hal Far, Malta, in mid-May 1951 for weapons training, returning to the UK aboard HMS *Illustrious* in October 1951. The squadron flew its Firebrands aboard HMS *Eagle,* along with the Supermarine Attackers of No. 800 Squadron, on 4 March 1952 returning to Ford on 24 March. No. 827 Squadron embarked on HMS *Eagle* on a further three occasions during 1952, finally returning to RNAS Ford to be disbanded there on 3 December 1952.

It had been apparent from early in its development that the Firebrand was not the ideal design for carrier operations. Consequently, on 26 February 1944, Specification S28/43 was issued for a Firebrand with a redesigned wing and a much improved forward view for the pilot. Blackburn's design team set to work on a development of the Firebrand TF Mk V design. The new project, identified as the B-48 and unofficially called the "Firecrest" by the company, used the same basic construction as the Firebrand, but there the similiarity ended. The cockpit was raised and moved forward to improve the view over the nose, requiring the fuselage fuel tanks to be moved aft of the cockpit. The major change, however, was the completely new inverted gull mainplane. To provide the slow landing and take-off performance required for carrier operation, four Fowler type flaps were fitted to the wings and in addition retractable dive brakes were fitted in the upper and lower surfaces of the wings. Electrically operated trim tabs were fitted to all control surfaces. Each wing folded hydraulically in two places to facilitate storage on aircraft carriers.

The B-48 design showed considerable improvement over the Firebrand, and consequently an order was

B-48 (Firecrest) prototype VF172 on test flight from Brough. (B.Ae.)

placed for two prototypes. The first, RT651, was completed early in 1947 and transported to RAF Leconfield, where it flew for the first time on 1 April 1947 and, after proving its short airfield performance, was returned to the company's airfield at Brough on 9 April.

Group Captain C. J. P. Flood demonstrated RT651 at the Naval Air Command Display at RNAS Lee-on-Solent on 26 June 1947, and it made its next public appearance at the SBAC Show at Radlett on 9-11 September 1947.

Work on the second prototype, RT656, was abandoned but a third prototype, VF172, was completed. This differed from the first prototype in having power-boosted ailerons similar to those used on the Firebrand TF Mk 5A. As part of this modification the nine degree dihedral used on the outer wings of RT651 was reduced to only three degrees on VF172. The aircraft, flown by P. G. Lawrence, appeared at a Royal Aeronautical Society Garden Party at Brough on 21 August 1948 and was also demonstrated by Lawrence at the Farnborough SBAC Show the following month.

Consideration was given to a version of the B-48 powered by one of the new propeller-turbine engines that were becoming available in the early post-war period, but these plans came to nothing and further development of the B-48 project was abandoned, although the prototypes continued flying at Brough for some time afterwards.

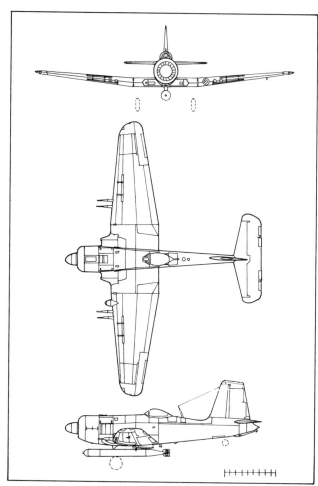

2 Firebrand TF Mk 5

Production

F Mk I DD804, DD810, DD815 (Prototypes). DK363 – DK371.

TF Mk II DK374 – DK385.

TF Mk III DK372, DK373 (Prototypes). DK386 – DK412.

TF Mk IV EK601 – EK638, EK653 – EK694, EK719 – EK740.

TF Mk 5/5A EK741 – EK748, EK764 – EK799, EK827 – EK850.

Firecrest RT651, RT656, VF172 (Prototypes).

CHAPTER THREE
De Havilland Sea Mosquito

Following the considerable success of the Mosquito in RAF service the Admiralty started, in 1943, to investigate the feasibility of using the type as a carrier-borne, long range, high speed strike aircraft. The main problem, inevitably, hinged around the ability to operate a large twin-engined aircraft from a carrier deck. Consequently a Mosquito FB Mk VI, LR359, was allotted for carrier trials and was modified by the introduction of a Fairey Barracuda type arrester hook mounted under the rear fuselage. To withstand the high loadings that occur during arrested landings, the rear fuselage was reinforced by the introduction of additional longerons. In an attempt to keep the weight to a minimum for the trials, the armament was deleted. The aircraft was, however, not fully navalised, lacking the folding wings necessary for carrier operation. LR359 was delivered to the RAE at Farnborough in January 1944, where preparations for the carrier trials commenced on 25 January 1944 with landings and take-offs being carried out on a dummy deck marked out on the runway. This allowed the operating techniques to be refined prior to the actual carrier trials.

A second semi-navalised Mosquito FB Mk VI, LR387, arrived at Farnborough on 1 March 1944 to act as a back-up aircraft for the trials. Both aircraft were flown to RNAS Arbroath on 18 March, where LR387 was used for arrester gear proofing tests using arrester wires installed across the runway. Afterwards, Airfield Dummy Deck Landings (ADDLs) were carried out at East Haven using LR359, and following the satisfactory completion of these trials the aircraft were flown to

Navalised Mosquito FB Mk VI, LR359, preparing for take-off during carrier trials on HMS *Indefatigable* in May 1944.

Machrihanish on 23 March. Two days later, Lt. E. M. Brown took off in LR359 and shortly afterwards made a successful landing on HMS *Indefatigable,* the first deck landing by a twin engined aircraft. On this first day of the trials Lt. Brown made five successful landings and, although deck operations had proved rather easier than expected, the size of the Mosquito had required it to operate away from the deck centre line to ensure that the starboard wing was clear of the island.

The trials re-started the following day. The first two landings were uneventful, but on the third the arrester hook failed after picking up an arrester wire and it was Lt. Brown's fast reaction to the problem that averted a ditching and enabled him to take off and fly back to Machrihanish.

LR359 was returned to de Havilland at Hatfield during April 1944, while LR387 was delivered to the Service Trials Unit, commanded by Cdr. H. J. F. Lane, at Crail. Towards the end of April, LR359 returned to Farnborough following introduction of modifications which included the fitting of a tailwheel lock to reduce the swing on take-off, and a strengthened arrester hook, and LR387 joined LR359 at Farnborough a few days later.

On 6 May LR359, flown by Lt. Brown, and LR387, flown by Cdr. Lane, were delivered to Crail. The following day Lt. Brown flew LR387 to Arbroath for a further series of arrester proofing trials and from there, on 8 May, to Machrihanish, calling at East Haven en route for ADDLs. The following day he flew LR387 to HMS *Indefatigable* to continue the deck landing trials. On this occasion LR359 was held as the back-up aircraft and the flying of LR387 was shared by Lt. Brown and Cdr. Lane.

A total of 24 landings were made during the carrier trials and these showed that the main problem was the swing on take-off.

The first of the semi-navalised Mosquitos, LR359, was written off in an accident at Arbroath on 9 November 1944, and about this time the second, LR387, was returned to de Havilland to be converted to full naval standard by the introduction of manually folded outer wings and the American ASH radar which necessitated a thimble shaped radome in the nose. In effect, therefore, LR387 became the prototype of the Sea Mosquito TR Mk 33, designed to meet Specification

N.15/44 which was basically a navalised Mosquito FB
Mk VI. In addition to the prototype, two pre-production
development aircraft, TS444 and TS449, were built at
de Havilland's factory at Leavesden and these were
followed by a production order for 50 aircraft. An
order was also placed for a further 50 but this was later
cancelled. The first production aircraft, TW227, made
its initial flight at Leavesden on 10 November 1945.
Early production aircraft, like the two pre-production
machines, were not fitted with folding wings. It was not
until the 14th production aircraft, TW241, that wing
folding was introduced as standard. There were,
however, other improvements introduced into the
production TR Mk 33s, and these included a bulkhead
fitted into the fuselage at the arrester hook attachment
points and a new long stroke Lockheed pneumatic
undercarriage to reduce the bounce when landing on
carriers. The armament consisted of four 20mm cannons
in the nose along with the facility to carry two 500lb
bombs in the bomb bay or an 18in torpedo, a mine or a
2000lb bomb externally under the fuselage. Additionally, bombs, rocket projectiles or fuel drop tanks could
be carried under the wings. To permit carrier operations at maximum weight, the aircraft was cleared to
use rocket assisted take-off gear (RATOG). This had
been tested by Lt. E. M. Brown during trials at
Farnborough using the second of the pre-production
aircraft, TS449, during the summer of 1946.

Production Sea Mosquito TR Mk 33s were delivered
to No. 778 Squadron, the Service Trials Unit at Ford, in
April 1946 and, later the same month, 12 Mosquito FB
Mk VIs being used for familiarisation by the first line
No. 811 Squadron at Ford were replaced by an equal
number of Sea Mosquito TR Mk 33s. Although the
aircraft had been cleared for carrier operation, there is

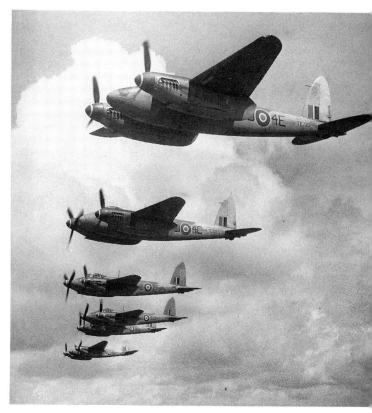

A formation of six Mosquito FB Mk VIs, including TE720/FD-4E
and TE721/FD-4C, of No. 811 Squadron in June 1946.

Sea Mosquito TR Mk 33, TW256/593/LP, of No. 771 Squadron, Lee-
on-Solent. (RAF Museum)

First production Sea Mosquito TR Mk 37, VT724. (B.Ae.)

Sea Mosquito TT Mk 39, PF606.

no record of No. 811 Squadron's Sea Mosquitos ever going to sea. First line operation of the Sea Mosquito was fairly short, No. 811 Squadron reducing to half its initial strength when it moved to RNAS Brawdy in December 1946, and after a final move to RNAS Eglinton in March 1947 it disbanded there on 1 July 1947.

This was not the end of the Sea Mosquito TR Mk 33 in Fleet Air Arm service as a number of them continued to operate with second line squadrons, including Nos. 703, 739, 751, 771, 778 and 790. The last squadron to use the Sea Mosquito was No. 751, the Radio Warfare Unit which operated at RAF Watton and phased its last aircraft out of service in June 1953.

The next development of the Sea Mosquito was the TR Mk 37, which saw the introduction of the British ASV Mk XIII Radar. The larger scanner of the British radar required a complete redesign of the nose, but otherwise the aircraft was externally identical to the earlier version. A production Sea Mosquito TR Mk 33, TW240, was used as the prototype for the TR Mk 37, and subsequently 14 production aircraft were built at de Havilland's factory at Chester. These were to serve with the second line squadrons Nos. 703 and 771.

The final version of the Sea Mosquito, the TT Mk 39, without any doubt the ugliest version of the Mosquito, was a special conversion of the Mosquito B Mk XVI to meet Specification Q.19/45 for a target tug to replace the Miles Monitor TT Mk II. The design and conversion work was undertaken by General Aircraft Ltd at Hanworth and the project received the designation GAL 59. The conversion provided for two new crew members, a camera operator located in an extended and glazed nose and an observer positioned in a dorsal cupola located just aft of the wing trailing edge. General Aircraft Ltd produced two prototypes, ML995 and PF569, and 24 production aircraft. In addition to its basic target towing role, the aircraft could also be used for radar calibration duties. The first aircraft were delivered to No. 703 Squadron, the Naval Air Sea Welfare Development Unit at RAF Thorney Island, in October 1948 and deliveries to the main user of the type, No. 728 Squadron at Hal Far, Malta, commenced in March 1949. Sea Mosquito TT Mk 39s were also operated by No. 771 Squadron at Lee-on-Solent for two years from January 1950, but it was No. 728 Squadron that was the final operator of the TT Mk 39s, eventually phasing them out of service in May 1952 when they were replaced by the Short Sturgeon TT Mk 2.

Production

TR Mk 33 TS444, TS449, TW227 – TW257, TW277 – TW295.

TR Mk 37 VT724 – VT729.

TT Mk 39 ML936, ML956, ML974, ML980, ML995, MM112, MM117, MM144, MM156, MM177, PF385, PF439, PF445, PF449, PF452, PF481 – PF483, PF489, PF512, PF513, PF522, PF557, PF558, PF560 – PF562, PF568 – PF571, PF575, PF576, PF599, PF605, PF606, PF609, PF612, PF616, RV295.

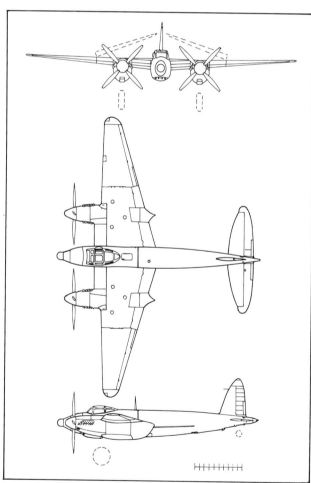

3   Sea Mosquito TR Mk 33

CHAPTER FOUR
Miles Monitor

Specification Q.9/42 was issued in 1942 for a twin-engined target tug for the RAF. The aircraft had to have the capability of towing a target at 300 mph at 20,000ft, and to be operated by a crew of two. The Air Ministry considered the requirement to be so urgent that they suggested that a Bristol Beaufighter wing and undercarriage should be incorporated to save time in both design and production phases of the project.

The design team at Miles Aircraft Ltd, under the leadership of G. M. Miles, prepared their initial design study, the M 33, to meet the specification. Although they were unable to use the Beaufighter wing, using a wooden one of their own design instead, they did include its undercarriage. The fuselage was an all metal structure. Towards the end of 1942 it was decided that the Royal Navy's requirements for a target towing aircraft were more urgent than the RAF's, and consequently work was put in hand to re-design the M33 accordingly. This involved towing targets for ship-to-air and ground-to-air gunnery practice as well as simulat-

Monitor TT Mk II, NP407.

ing dive-bombing attacks on warships, requiring the introduction of dive brakes. The design of the M33, by then named Monitor, was completed early in 1943 and the prototype, NF900, was flown for the first time by Miles test pilot T. Rose on 5 April 1944. NF900 proved to be very pleasant to fly, but during the flight development programme, whilst being flown by H. V. Kennedy, a problem developed in the undercarriage which prevented one main undercarriage leg being lowered. As a result, the pilot had to make a wheels-up landing, successfully accomplishing this with minimal damage to the aircraft. It was soon repaired and returned to the flying programme. Unfortunately, a few months later when being flown by an A&AEE crew from Boscombe Down, the aircraft burst into

flames when on its final approach to Woodley aerodrome and crashed. The pilot managed to bale out, but the observer failed to clear the aircraft and was killed.

The original project to meet the RAF requirement was designated Monitor TT Mk I, the Royal Naval aircraft becoming the TT Mk II. At an early stage in the development programme a production order was placed for 600 TT Mk IIs but, with the approach of the end of the war, this order was first reduced to 200 aircraft, then to 50 and, by the time 20 had been built, the contract was finally cancelled. Ten were delivered to the Fleet Air Arm, but none was to see squadron service and it is assumed that the aircraft went straight into store to be subsequently scrapped. The remaining ten were broken up at Woodley.

Production

Prototype NF900.
TT Mk II NP406 – NP425.

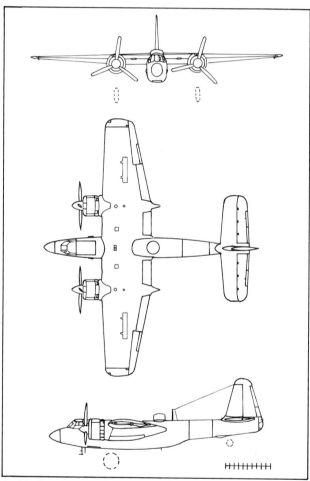

4   Monitor TT Mk II

CHAPTER FIVE
Fairey Barracuda (Mk V)

Barracuda TR Mk V, RK571/800/D, of HMS *Illustrious'* Ship's Flight. (via G. A. Jenks)

Barracuda TR Mk V, RK558/325/LP, of No. 783 Squadron, Lee-on-Solent.

The Barracuda was designed by Fairey Aviation Ltd during 1938 to meet Specification S.24/37 for a torpedo/dive-bomber/reconnaissance aircraft as a replacement for the Fairey Albacore. Initially identified by Fairey as the Type 100, the Barracuda was a shoulder-wing monoplane of conventional stressed-skin construction, initially powered by a Rolls Royce Merlin 30 engine, later versions being powered by the more powerful Merlin 32. The Barracuda was operated by a crew of three. Hydraulically operated Youngman flaps were fitted below the wing trailing edge, which were used as dive-brakes as well as for take-off and landing.

A total of over 2,500 Barracudas were built during World War Two and although the aircraft was generally disliked by the aircrew, they were used to good effect in 1944 during attacks on the *Tirpitz,* which was lying crippled in Kaafjord, North Norway, after being damaged in an attack by midget submarines. At about the same time Barracudas were also heavily involved in

raids against Japanese targets in Sumatra.

Towards the end of the war a major redesign of the Barracuda was undertaken to provide an interim aircraft for use in the war against Japan, until the Fairey Spearfish became available. This development resulted in the Barracuda TR Mk V, incorporating the 2,020hp Rolls Royce Griffon 37 driving a Rotol four-bladed, constant-speed propeller. The wing-span was increased by four feet and additional fuel tanks were introduced. The size of the fin and rudder was significantly increased and the airframe structure was strengthened to provide a greater margin of safety during the pull-out from a dive. The crew was reduced to two, a pilot and an observer telegraphist who was also navigator and radar operator. The radar scanner was fitted in a radome mounted in the leading edge of the port wing. The armament was a single forward firing machine gun, and additionally a variety of bombs, depth charges or mines could be carried under the wings and a single 18in torpedo under the fuselage.

The first prototype Barracuda TR Mk V was a converted Mk II, P9976, although not to full production standard, being fitted with a 1,850hp Griffon VII engine. This aircraft, which was converted at Stockport, flew for the first time as a Mk V at Ringway on 16 November 1944. Shortly afterwards a production contract was placed with Fairey for 140, although initially a small quantity of Barracuda Mk IIs were converted to Mk Vs, retaining their original rudder but with an extended fin. The first new build Mk V, RK530, flew for the first time on 22 November 1945, by which time the Second World War had come to an end and consequently the order for the Barracuda Mk V was reduced to 30. Towards the end of the production run a small number of aircraft were fitted with a new tall pointed rudder.

Deliveries to the Fleet Air Arm started with service trials carried out by No. 778 Squadron commanded by Lt. Cdr. R. H. P. Carver DSC, at RNAS Ford between September 1946 and July 1947. The only squadron to actually operate the Barracuda TR Mk V was No. 783 Squadron commanded by Lt. Cdr. K. C. Winstanley at RNAS Lee-on-Solent, which provided support for the Naval Air Signal School at Seafield Park. This squadron received six aircraft in December 1947 as an interim measure until a flying classroom version of the Anson became available. The Barracudas were phased out of service in October 1948.

The only other known users of the Barracuda TR Mk Vs were the Ship's Flights of HMS *Illustrious,* which used RK571 during 1947 and 1948, and HMS *Implacable,* which used RK568 between February and June 1948.

Production

TR Mk V Conversions P9976 (Prototype).
DT845, PM940, PM941, PM944, LS479, LS486.
New Build RK530 – RK542, RK558 – RK574.

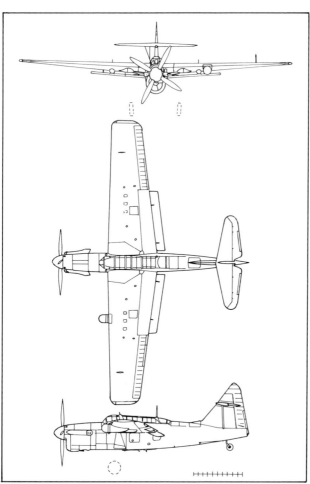

5 Barracuda TR Mk V

CHAPTER SIX

De Havilland Sea Hornet

Following the considerable success of their DH98 Mosquito, the de Havilland design team started work in 1942 on the design of a twin engined long range fighter, as a private venture, utilising wherever possible technology developed for the Mosquito. Identified as the DH103, the new project was a very streamlined aircraft to be powered by the new 2030hp Rolls Royce Merlin 130 and 131 engines. These engines were handed, turning the propellers in opposite directions and consequently cancelling the adverse effect of engine torque. Good progress was made with the design and a full scale mock-up was completed in January 1943. In June of that year the Ministry of Aircraft Production issued Specification F.12/43, which was written around the DH103 project, and at about the same time two prototypes were ordered.

The construction of the DH103 was an advance on the Mosquito, using a wood/light alloy bonded structure rather than the Mosquito's ply/balsa wood bonded

Three Sea Hornet FR Mk 20s of No. 801 Squadron, Ford, including TT196/150/FD, flying over HMS *Implacable* in October 1947.
(RAF Museum)

construction. The pilot was located in a cockpit in the fuselage nose and a clear view canopy was provided, giving him excellent all round visibility. The armament consisted of four 20mm Hispano cannons mounted under the nose, with provision for two 1000lb bombs or eight 60lb rocket projectiles to be mounted under the outer wings.

By the time the first prototype, RR915, made its maiden flight at Hatfield in the hands of Geoffrey de Havilland on 28 July 1944, the type had been named "Hornet". The second prototype, RR919, joined the test flight programme shortly afterwards. Before the end of 1944 an order had been placed for 60 Hornet F Mk Is for the RAF and the first of these, PX210, was completed early in 1945.

From the outset of the design, the de Havilland designers had realised that the Hornet was very suitable for development into a carrier-borne fighter. The contra-rotating propellers, the excellent forward view for the pilot and the long range of the aircraft were ideal for a naval fighter. The de Havilland proposals for the development of a naval version of the Hornet were

Sea Hornet FR Mk 20, VZ708/456/C, of No. 801 Squadron, HMS *Implacable* in March 1950. (RAF Museum)

accepted by the Ministry of Aircraft Production, and late in 1944 Specification N.5/44 was issued to cover the development.

To speed development, three production Hornet F Mk Is, PX212, PX214 and PX219, were allocated to the naval programme. However, because of de Havilland's commitments, the design and conversion work was subcontracted to the Heston Aircraft Company. The modifications necessary to meet the naval requirements included the introduction of a Lockheed hydraulic wing folding system, a V-frame arrester hook, catapult pickup points and a new pneumatic undercarriage capable of withstanding the high landing loads that occur during carrier operations.

The first two prototypes were only partially converted to the naval requirement, not being fitted with folding wings, and it was the third prototype, PX219, which was the first fully navalised Hornet. The first aircraft, PX212, made its maiden flight at Heston on 19 April 1945, by which time the type had been named the Sea Hornet. Later, two additional Hornet F Mk Is, PX230 and PX239, were transferred to Heston to be converted for the Sea Hornet development programme. Following manufacturer's trials, PX212 was delivered to the RAE at Farnborough for trials on the dummy carrier deck to assess its handling before commencing deck landing trials proper. The trials at Farnborough were carried out by the senior naval test pilot Lt. Eric (Winkle) Brown who, although very impressed with the flying characteristics of the Sea Hornet, did encounter some problems during the deck landings, which had to be resolved before the Sea Hornet went to sea. The lateral control at low speed had proved to be inadequate and the pilot's flap lever and throttle quadrant needed modifications to suit carrier operations. These problems were soon corrected by de Havilland's at Hatfield and on 4 August 1945 Lt. Brown flew PX212 to RNAS Arbroath, where he carried out a programme of Aerodrome Dummy Deck Landings (ADDLs). Two days later he flew the aircraft to RNAS Ayr, and on 10 August he landed PX212 aboard HMS *Ocean* at the commencement of the initial carrier trials and, after the successful completion of these on 11 August, PX212

was flown back to Farnborough. Shortly after the trials a production order for 80 Sea Hornet F Mk 20s was placed with de Havilland's.

The first production Sea Hornet F Mk 20, TT186, flew for the first time at Hatfield on 13 August 1946 and took part in the SBAC Show at Radlett the following month. The production aircraft differed from the prototypes in being fitted with slotted flaps to further improve their deck landing characteristics. TT186 was delivered to No. 703 Squadron at RAF Thorney Island during October 1946. Prior to this, however, in March 1946, the Service and Carrier Trials Unit, No. 779 Squadron at RNAS Ford, had received a number of early development Sea Hornets to start the service trials programme. The squadron continued to operate Sea Hornets until July 1948, when the service trials unit task was taken over by No. 703 Squadron, which by that time had moved to RNAS Lee-on-Solent.

In October 1947 one of the production aircraft, TT213, was allotted to de Havilland's at Hatfield for tropical trials. It was subsequently re-allotted to the Royal Australian Air Force at Point Cook, Victoria, Australia, for trials. After being dismantled and packed by No. 47 Maintenance Unit at RAF Sealand, it was shipped to Australia in April 1948. On arrival in Australia the serial number A83-1 was allocated but does not appear to have been applied to the aircraft, which thus carried its British serial number throughout the trials. The trials, carried out at Oodnadatta, Laverton and Woomera, were successfully completed by the end of 1950 and the aircraft was returned to the United Kingdom early in 1951.

There was only one first line Sea Hornet F Mk 20 squadron and that was No. 801, which had reformed at RNAS Ford with six Sea Hornet F Mk 20s on 1 July 1947, commanded by Lt. Cdr. D. B. Law DSC. The squadron was brought up to its full strength of 12 in October when it joined with No. 813 Squadron to form the 1st Carrier Air Group. Following a period of working up at RNAS Ford, deck landing training commenced on 19 November 1947 when a flight of four aircraft embarked on HMS *Implacable* for four days' intensive training. In January 1948 the squadron transferred from Ford to RNAS Culdrose, and on 5 March embarked on HMS *Implacable* for several cruises in Home and Mediterranean waters. At the end of 1950 the squadron embarked on HMS *Indomitable* for two weeks, shortly before returning to *Implacable* for the squadron's last sea-borne operations with the Sea Hornet. On its return to Lee-on-Solent in March 1951, the squadron re-equipped with twelve Sea Fury FB Mk 11s.

In May 1948, on the direction of the Admiralty, a flying display team was formed and identified as No. 806 Squadron, to take part in the Air Exposition in August 1948 celebrating the opening of Idlewild Airport in New York. The team comprised two Sea Fury FB Mk 11s, two Sea Hornet F Mk 20s, plus one in reserve, and a Sea Vampire F Mk 20. The aircraft were to be flown by a team of four specially selected pilots, including the commanding officer Lt. Cdr. D. B. Law, who had been transferred from No. 801 Squadron to set up the team. After practising and refining the display sequence at RNAS Eglinton, the squadron embarked on HMCS *Magnificent,* which was to take them to Canada. They arrived at the Naval Air Station at Dartmouth, Nova Scotia, on 2 June 1948 with flying re-commencing on 4 June. The following day one of the Sea Hornets, VR845 flown by Lt. N. D. Fisher, crashed into Halifax Harbour, killing the pilot. The squadron had put together a most impressive display sequence by the time the first public display was given at NAS Dartmouth on 21 July, with the commanding officer Lt. Cdr. Law and Lt. I. H. F. Martin flying the Sea Hornets. Much of the Sea Hornets' display, including loops, rolls and inverted flying, was carried out with one engine feathered and, to bring the display to a conclusion, Lt. Martin completed two consecutive loops with both propellers feathered. Following a very successful tour of Canada and the USA, the squadron returned to the United Kingdom and was disbanded on 25 September 1948.

The Naval Air Fighter School, No. 736 Squadron at RNAS Culdrose which at the time was operating 50 Sea Furies, took a small number of Sea Hornet F Mk 20s on charge in February 1950 and operated them until June 1951. On 1 May 1950 this very large squadron was split up and part of it formed No. 738 Squadron, which was equipped with a variety of aircraft including Seafires, Sea Furies, Firebrands and a small number of Sea Hornets. The Sea Hornets remained with the Squadron until August 1951.

The eighth production Sea Hornet F Mk 20, TT193, was retained by de Havilland at Hatfield until it was sent to the Royal Canadian Air Force Station at

Edmonton, Alberta, for cold weather trials. On completion of the trials, the aircraft was purchased by the survey company Spartan Air Services Ltd of Ottawa on 1 July 1950.

The final production Sea Hornet F Mk 20, WE242, was delivered on 12 June 1951, by which time all of the F Mk 20s had been relegated to second line duties, ending their service with the Fleet Requirement Unit, No. 728 Squadron, at RNAS Hal Far, Malta, where they were finally phased out of service in February 1957.

In 1945 the Royal Navy had a requirement for a night fighter, and to meet this requirement it was decided to modify the Sea Hornet, Specification N.21/45 being issued to cover the modification. The Heston Aircraft Company was again given the task of converting one of the Sea Hornet F Mk 20 prototypes, PX230 into the night fighter prototype, now identified as the Sea Hornet NF Mk 21. The changes included the introduction of a radome in the nose to accommodate the ASH radar scanner, and an additional cockpit in the fuselage just aft of the wing trailing edge for an observer. The engines were changed to the slightly lower powered Merlin 133/134s, and flame damping exhaust manifolds were introduced.

The first prototype NF Mk 21, PX230, made its maiden flight on 9 July 1946 and another F Mk 20 prototype, PX239, became the second NF Mk 21 prototype. This was the first fully navalised NF Mk 21 as, unlike PX230, it was equipped with folding wings. During trials at the A&AEE Boscombe Down, when being flown by Lt. Cdr. K. Hickson on 12 May 1947, PX230 had an engine break away from its mountings and the airframe started to disintegrate. The fuselage eventually broke in half at the observer's cockpit, which was fortunately empty. At this point, almost incredibly, the pilot succeeded in bailing out and, although at low altitude, made his escape uninjured.

Carrier trials using PX239 commenced aboard HMS *Illustrious* on 25 October 1948 and included both day and night landings, which were completed without incident.

A contract for 78 Sea Hornet NF Mk 21s was placed with de Havilland and the first production aircraft, VV430, made its first flight at Hatfield on 24 March 1948. Towards the end of 1948, after the first 18 production NF Mk 21s had been completed, production was transferred to de Havilland's newly acquired factory at Broughton, near Chester, where the last aircraft was completed in November 1950.

The Sea Hornet NF Mk 21 entered first line service on

Flight of Sea Hornet NF Mk 21s, including VV437/484/CW of No. 809 Squadron, Culdrose, in 1949.

Sea Hornet PR Mk 22, VZ664/451/CW, of No. 738 Squadron, Culdrose, 1951. (A. E. Hughes)

20 January 1949 when No. 809 Squadron, commanded by Major J. O. Armour RM, reformed at RNAS Culdrose with four NF Mk 21s; the complement being later increased to eight aircraft. Deck landing training commenced in January 1950 when a detachment of three of 809 Squadron's aircraft were flown aboard HMS *Illustrious* for a few days. This was followed by the squadron flying aboard HMS *Vengeance* on 12 May 1950 to become part of the 15th Carrier Air Group, and during the next twelve months spending a further two short spells in HMS *Vengeance*. On 16 October 1951, four of the squadron's Sea Hornets flew non-stop in formation from Gibraltar to Lee-on-Solent, a distance of 1040 miles, in 3 hours 10 minutes. The following

month a single aircraft flying the same route reduced the time taken to 2 hours 45 minutes. In January 1952 the squadron moved to RNAS Hal Far in Malta, returning home to Culdrose on 24 March 1952. The squadron then spent two weeks aboard HMS *Indomitable* in June 1952, the only time it went to sea during 1952. In January 1953, No. 809 Squadron flew aboard HMS *Eagle* and was to remain as part of *Eagle's* air group until disbanding on 10 May 1954.

A number of Sea Hornet NF Mk 21s continued to operate on second line duties, primarily with the civilian operated Fleet Requirements Units operated by Airwork Ltd at Hurn, Brawdy and St David's. The unit at St David's also provided a Heavy Twin Conversion Course for Royal Navy pilots using Sea Hornets and, for a short while, Sea Mosquitos. Sea Hornet NF Mk 21s continued to be used by these units until October 1955, when they were replaced by Sea Furies, Attackers and Sea Venoms.

The final version of the Sea Hornet was the unarmed PR Mk 22, which was a photo-reconnaissance version of the F Mk 20 with provision for two standard F.52 cameras for daytime use and a single K.19B for night operations. Although the aircraft was still equipped to carry underwing armament, the cannons in the nose were removed and gun ports faired over. The second production F Mk 20, TT187, was converted at Hatfield in 1948 to act as prototype of the PR version, and a total of 23 production aircraft were built. Three of these, VZ658, VZ659 and VZ660, were attached to No. 801 Squadron in December 1949 until March 1950. Also a single PR Mk 22, VW931, was used by No. 1833 Squadron, Royal Naval Volunteer Reserve, at RNAS Bramcote from June 1951 until February 1952. Most of the PR Mk 22s were used by second line squadrons, but a number of the late production aircraft were never used, being put into long term storage immediately after delivery, and remaining so until they were scrapped along with most of the other Sea Hornets in 1957.

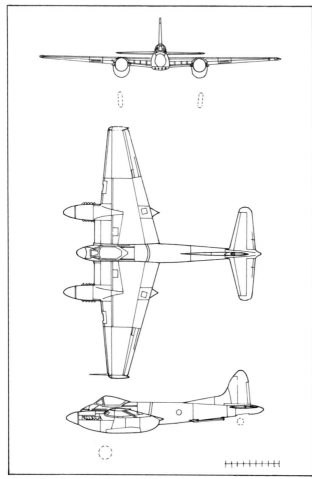

6   Sea Hornet F Mk 20

Production

F Mk 20 PX212, PX214, PX219, PX230, PX239 (Prototypes).
TT186 – TT213, TT247, TT248, VR837 – VR864, VR891, VR892, VZ707 – VZ715, WE235 – WE242.
NF Mk 21 VV430 – VV441, VW945 – VW980, VX245 – VX252, VZ671 – VZ682, VZ690 – VZ699.
PR Mk 22 VW930 – VW939, VZ655 – VZ664, WE 245 – WE247.

CHAPTER SEVEN

Hawker Sea Fury

The Sea Fury was the culmination of a continuous line of design and development that had started in 1937 with the issue of Specification F.18/37 and which resulted in the Tornado and Typhoon. Although the Tornado project was abandoned, the Typhoon, after a period of intensive development, proved its worth as a low level, ground attack aircraft. A major redesign of the

Typhoon led to the Tempest, which became a highly successful fighter and ground attack aircraft.

During the early development of the Tempest, Sydney Camm proposed a light-weight version, which was basically a Tempest without a separate wing centre section and with a fuselage of stressed-skin, semi-monocoque construction. Following discussions between the Air Ministry and Hawkers, Specification F.6/42 covering the project was issued late in 1942. This was further refined and re-issued as F.2/43 early in 1943.

Early production Sea Fury F Mk X, TF898, during carrier trials on HMS *Illustrious* in March 1947. (B.Ae.)

Two Sea Fury F Mk Xs, TF915/161/JA and TF945/162/JA, of No. 767 Squadron, Stretton. (J. Thomason)

At about this time the Royal Navy had a requirement for a fleet fighter and Sydney Camm suggested that his F.2/43 project fitted with the up-rated Centaurus XII engine could also satisfy the Naval Specification N.7/43. However, with Hawker's heavy production commitments, it was decided that they would only be responsible for the RAF version with Boulton-Paul at Wolverhampton as the main contractor for the naval version.

Six prototypes of the RAF version were ordered in 1943, two powered by Bristol Centaurus XXIIs, two by Rolls Royce Griffons, one by a Bristol Centaurus XII, and one to be used as a structural test specimen.

The production standard of the naval aircraft was covered by Specification N.22/43, and early in 1944 a contract was placed with Hawker for the manufacture of three prototypes: one of these was subsequently sub-contracted to Boulton-Paul. In April 1944, a contract was placed for the production of 200 F.2/43 for the RAF and 200 N.22/43 for the Royal Navy, production of 100 of the naval aircraft being allotted to Boulton-Paul.

The first prototype F.2/43, NX798, powered by a rigidly mounted Bristol Centaurus XII engine driving a Rotol four-bladed airscrew, made its maiden flight at Langley on 1 September 1944 in the hands of Philips Lucas. The second prototype, LA610, powered by a Rolls Royce Griffon 85 engine with a Rotol contra-rotating six-bladed airscrew, first flew on 27 November 1944. Before the end of 1944, the RAF aircraft had

been named Fury and the Royal Navy version named Sea Fury.

With the success of the early jet fighters and the end of the war in sight, the contract for Furies for the RAF was cancelled in January 1945. The Royal Navy still had serious doubts about the suitability of jet powered aircraft for carrier operations, and it was therefore decided to continue with the Sea Fury, although the initial production order was reduced to 100 aircraft. This resulted in the remaining production work being concentrated at the Hawker factory and Boulton-Paul losing its entire production order.

The first Sea Fury prototype, SR661, powered by a Centaurus XII engine driving a Rotol four-bladed airscrew, flew for the first time on 21 February 1945. This aircraft was virtually to full naval standard, lacking only the folding wings necessary for production aircraft. Problems with the Centaurus engine resulted in both SR661 and the Fury prototype NX798 making a number of forced landings on or around Hawker's airfield at Langley. The problem with the engine was later identified as a defective lubrication system.

SR661 was delivered to the Royal Aircraft Establishment (RAE) at Farnborough in May 1945 for an assessment of rudder control for carrier operation. With the strong swing to starboard on take-off rudder control was critical, and the trials showed that there was a tendency to over-correct, setting up a slight swing to port. The Sea Fury also proved to be very unstable directionally during the landing run. The rigid mountings of the Centaurus engine produced severe vibration when the engine was running at low speed and any sudden application of power resulted in the airscrew overspeeding.

Hawker modified SR661, replacing the original four-bladed airscrew with one with five blades, and introducing a redesigned fin and rudder of increased area. Early in July 1945, the aircraft returned to Farnborough for arrested landing trials. However, because of poor arrester hook damping, the trials were not a success and the arrester hook mechanism had to be modified before the trials were satisfactorily completed by the end of July. Early in August SR661 was delivered to the A&AEE at Boscombe Down to prepare for carrier trials. On 10 August it was flown on board HMS *Ocean* for initial carrier trials, and these trials were completed without incident.

The second Sea Fury prototype, SR666, differed from the first in being powered by a Centaurus XV engine with a Rotol five-bladed airscrew, and was fully navalised with folding wings and armament. It flew for the first time on 12 October 1945, and in mid 1946 was delivered to Farnborough for acceleration tests using the RAE catapult. Following the successful completion of these, carrier trials were carried out on HMS *Victorious* by the Service Trials Unit. During the trials there was nearly a serious accident when the arrester hook was snapped off during a landing. It was only the prompt action of the pilot that enabled the aircraft to take off again and land ashore rather than end as an undignified wreck in the carrier's crash barrier.

The final Sea Fury prototype, VB857, flew for the first time on 31 January 1946, and was powered by a Centaurus XXII fitted with a modified lubrication system and mounted on a flexible Dynafocal mounting. The manufacture of this aircraft had been started by Boulton-Paul but it was completed by Hawker.

Production Sea Furies, designated the F Mk X, started to come off the Langley production line in 1946 with the first aircraft, TF895, being taken on its maiden flight on 7 September by Hawker test pilot E. S. Morrell. The first few production aircraft were fitted with four-bladed Rotol airscrews, but these were soon superseded by the more satisfactory five-bladed airscrew.

Most of the early production Sea Fury F Mk X aircraft were used for the development trials, including several that were allotted to the Bristol Engine Company for Centaurus development. During March 1947 three aircraft, TF898, TF899 and TF900, were flown aboard HMS *Illustrious* for a programme of intensive deck landing trials. On 24 March TF899 crashed into the barrier following the failure of the arrester hook and was unserviceable for the remainder of the trials.

Two aircraft, TF902 and TF908, commenced a programme of intensive flying trials on 4 June 1947 at the A&AEE Boscombe Down. The aim was to achieve 100 flying hours per aircraft under typical service conditions. Although the programme was disrupted by long delays because of difficulty in obtaining spares, the A&AEE report was complimentary to the aircraft, stating that: "In general, the aircraft was found to be robust, easy to maintain, and was well liked by air and ground crews. It is likely to be popular in Service use", and in addition: "All pilots liked the Sea Fury, considering it exceptionally pleasant to fly and a good fighter aircraft".

It soon became apparent that the Royal Navy was in need of a fighter bomber. It was considered that the later marks of Seafire would cover the basic fighter

requirement, while the Sea Fury had sufficient development potential for the fighter bomber role. Consequently, the F Mk X was replaced on the production line, after only 50 aircraft had been built, by the Sea Fury FB Mk 11. In the event only relatively minor modifications proved necessary to convert the Sea Fury to the fighter bomber role. The changes were primarily the ability to carry a variety of external stores under the wings, ranging from 2 x 1000lb bombs to 12 x 60lb rocket projectiles, and provision for rocket assisted take-off (RATOG). Trials with the external stores had been carried out using SR666 and TF923. The third prototype, VB857, had been used for RATOG trials at the RAE Farnborough.

Deliveries of Sea Fury F Mk Xs to the Fleet Air Arm commenced in 1947 with the first aircraft being delivered during February to No. 778 Squadron, where they were tested by the squadron's Intensive Flying Development Flight at Royal Naval Air Station (RNAS), Ford, until July. Sea Furies were also delivered in May 1947 to No. 787 Squadron, which was the Naval Air Fighting Development Squadron of the Central Fighter Establishment at RAF West Raynham. Rather strangely, the first front line squadron to operate the Sea Fury was No. 803, Royal Canadian Navy, at RNAS Eglinton in Northern Ireland, which received 13 Sea Fury F Mk Xs during August 1947 to replace its Seafire F Mk XVs. However, it was a Royal Navy squadron, No. 807, which, although receiving its Sea Furies a month after No. 803, was the first to go to sea when it embarked aboard HMS *Implacable* on 4 May 1948, forming part of No. 17 Carrier Air Group (CAG). No. 803 flew its Sea Furies aboard HMCS *Magnificent* on 20 May, forming part of No. 19 CAG.

Only two other first line squadrons were to receive Sea Fury F Mk X aircraft, No. 802 re-equipped with

Sea Fury FB Mk 11, VX669/105/Q, of No. 802 Squadron on HMS *Vengeance*.

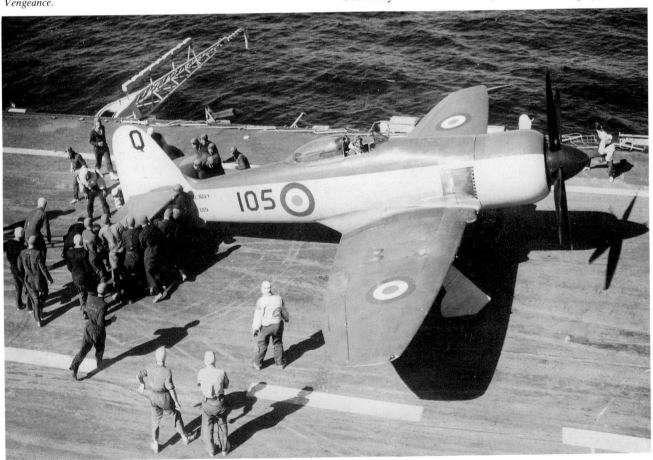

Sea Fury FB Mk 11, VX656/151/BR, of No. 1833 RNVR Squadron, Bramcote, landing at RNAS Stretton in 1955.

Sea Furies in April 1948 at RNAS Lee-on-Solent, and the Royal Australian Navy Squadron No. 805 re-equipped with a mixture of Sea Fury F Mk X and FB Mk 11 aircraft at RNAS Eglinton in August 1948. The Royal Navy retained the Sea Fury F Mk X in front line service for only a short time and by the end of 1948 they had been replaced in Nos. 802 and 807 Squadrons by the Sea Fury FB Mk II. The Australian and Canadian squadrons used the Sea Fury F Mk X rather longer, with the re-equipment of No. 805 Squadron being completed in February 1949, and No. 803 in February 1950. This was, however, not the end of the Sea Fury F Mk X, as it continued to be used by second line squadrons for several years. No. 767 Squadron, a Deck Landing Control Officer training squadron at RNAS Stretton, was apparently the last to use the type, phasing them out of service in early 1954.

Late in 1947, two Sea Furies, TF901 and TF909, were delivered to Canada. TF901 was used for demonstrations and evaluation trials by the RCN, while TF909 carried out cold weather trials with the Winter Experimental Flight at the RCAF station Edmonton. When HMCS *Magnificent* sailed home to Canada, in addition to No. 19 CAG she also had on board the unique Royal Navy flying display team, equipped with two Sea Fury FB Mk 11s, two de Havilland Sea Hornet F Mk 20s and a single de Havilland Sea Vampire F Mk 20: the last on its way to carry out a tour of North America.

The flying display team had been given the title of No. 806 Squadron and, following a period at NAS Dartmouth where they refined their display of individual and formation aerobatics, they put on a very impressive display at Dartmouth on 21 July 1948. During this display, Lt. R. H. Reynolds DSC, while carrying out a series of vertical rolls in a Sea Fury, suffered overspeeding of the Centaurus engine which forced him to make a deadstick landing in front of the spectators. The squadron left for Floyd Bennett Field, New York, on 25 July 1948 in readiness for the week-long Idlewild Golden Jubilee Air Celebration, which commenced on 1 August. After the final rehearsal for the display, Lt. Reynolds landed and reported to the ground crew that the engine of his Sea Fury was running roughly. They quickly pointed out that the tips

of the propeller had been bent when he had touched the ground during one of his high speed very low level runs; a speciality of his. The tips of the propeller were trimmed back by about three inches overnight to prepare for the display the following day. The amazing displays of individual and formation aerobatics by No. 806 Squadron completely stole the show at Idlewild, where the pilots received enthusiastic applause from the crowd. After Idlewild, the squadron returned to Canada for a further nine displays before disbanding in September, following what could only be considered as a highly successful tour, and returning to the United Kingdom on 25 September 1948.

The Sea Fury FB Mk 11 started to enter first line service with the Royal Navy in February 1948 when No. 807 Squadron started replacing their F Mk Xs with FB Mk 11s. This programme was, however, not completed until December. Consequently, it was No. 802 Squadron which became the first to be fully equipped with the Sea Fury FB Mk 11, receiving all its aircraft by June 1948. No. 802 Squadron was selected to provide the official Royal Navy formation aerobatic team for 1951, the "Festival of Britain" year. A team of four was formed, led by the squadron's senior pilot Lt. E. M. Brown. The team gave excellent displays in Britain and on the Continent throughout summer 1951.

On 25 June 1950 the North Korean army invaded South Korea across the 38th Parallel. The United Nations Security Council reacted rapidly and a UN force was soon on its way to defend South Korea. The British Government put the British Eastern Fleet, headed by the carrier HMS *Triumph,* at the disposal of the UN commander and it was soon operating alongside the United States Navy off Korea.

HMS *Triumph* was replaced by HMS *Theseus* in October 1950, with 17 CAG on board, consisting of No. 807 Squadron with 21 Sea Fury FB Mk 11s and No. 810 Squadron with 12 Firefly FR Mk 5s. Despite the appalling weather conditions during the tour of operations, No. 807 Squadron flew 2320 operational sorties involving some 5600 flying hours. The success of the air activities from HMS *Theseus* was recognised by the award of the Boyd Trophy for 1950 to the 17 CAG. The Boyd Trophy is an annual award instituted in honour of Rear-Admiral Sir Denis Boyd for the year's outstanding feat of naval aviation.

HMS *Glory* with 14 CAG aboard, comprising No. 804 Squadron with 21 Sea Fury FB Mk 11s and No. 812 Squadron with 12 Firefly FR Mk 5s, replaced HMS *Theseus* in April 1951 and was to remain in the East for a year, with a break from Korean operations from 4 October 1951 to 16 January 1952 when the ship visited Australia. During this time, *Glory* was replaced by HMAS *Sydney* with 20 CAG aboard. This comprised Nos. 805 and 850 Squadrons with Sea Furies, and No. 817 Squadron with Fireflies. They operated off Korea until the return of HMS *Glory*. In this eighteen week period, 2366 sorties were flown with a loss of eight Sea Furies and three Fireflies. Carriers continued to operate off Korea in approximately six month cycles and HMS *Ocean*, a new ship to the area, replaced HMS *Glory* in April 1952. No. 802 Squadron with Sea Furies and No. 825 Squadron with Fireflies were embarked on HMS *Ocean*. Almost coincidental with their arrival was the appearance of Mig 15 fighters and it was during the ensuing summer that perhaps the most significant incident involving the Royal Navy occurred: the shooting down of a Mig 15 jet fighter by a piston engined Sea Fury. This happened on 9 August 1952, when a flight of four 802 Squadron Sea Furies, led by Lt. P. Carmichael flying WJ232/114/O, were intercepted by eight Mig 15s. In the ensuing dog fight, one Mig was seen to crash into a hillside and two others were badly damaged: a feat achieved without damage to any of the Sea Furies. The Sea Fury had proved itself to be a very successful fighter bomber in the difficult conditions existing in Korea.

The end of the Korean War and the introduction of the fighter bomber version of the Hawker Sea Hawk brought about the demise of the Sea Fury in first line service. It is interesting to note that No. 810, the last first line squadron to operate the Sea Fury, was only re-equipped with them in March 1954 specifically to undertake the strike role as part of the CAG aboard HMS *Centaur* during its operations in the Mediterranean. These aircraft only acted as a stopgap until the Westland Wyvern, which had been having considerable development problems, became available in sufficient quantities. No. 810 Squadron's Sea Furies flew home from RNAS Hal Far, Malta, on 22 March 1955 and the squadron was disbanded on arrival in the UK. Ultimately, a total of eight first line Royal Navy squadrons operated the Sea Fury FB Mk II and in all 537 were built for the Fleet Air Arm.

Sea Furies continued to be operated for some time by a few Royal Navy second line squadrons, including Nos. 736, 738, 769 and 799, but they had all but disappeared from the scene by the end of 1956.

In 1947 the Royal Navy Volunteer Reserve (RNVR) Air Branch was formed, comprising one anti-submarine and three fighter squadrons. This was subsequently expanded to a maximum of twelve squadrons. Initially the fighter squadrons were equipped with Seafires, but in 1950 three of the squadrons started to re-equip with Sea Furies although it was not until August 1951 that the first squadron, No. 1831, commanded by Lt. Cdr. R. I. Gilchrist MBE, and operating from RNAS Stretton, was fully re-equipped with Sea Furies. Presumably because of the demand for Sea Furies for the Korean War, the re-equipment of the other two squadrons, Nos. 1832 and 1833, with Sea Furies was deferred. As an interim measure No. 1833 Squadron had its Seafire Mk XVIIs replaced by Seafire Mk 47s while No. 1832 Squadron continued with a complement of six Sea Furies and nine Seafires. The conversion of No. 1832 Squadron to Sea Furies was finally completed in May 1953; No. 1833 Squadron converted late in 1954 and by January 1955 all six RNVR fighter squadrons had been equipped with Sea Furies. However, during 1955 Supermarine Attackers and the early marks of Hawker Sea Hawks were becoming available, after being phased out of service with the first line squadrons, and by the end of the year all the RNVR fighter squadrons had moved into the turbo-jet era. Subsequently only the civilian operated Fleet Requirements Unit (FRU) run by Airwork Ltd at Hurn continued to use Sea Furies along with a variety of other aircraft until April 1961.

In 1946 the Admiralty expressed an interest in a two-seat trainer version of the Sea Fury and before the end of the year had issued Specification N.11/46. The Hawker design team, in addition to producing the design to meet N.11/46, also produced the design for a similar training version of the Fury to meet a requirement for Iraq.

The Sea Fury trainer for the Royal Navy was identified as the T Mk 20 and was virtually a standard Sea Fury with an additional cockpit for the instructor. The forward cockpit was identical, as regards location, equipment and layout, to that of the FB Mk 11. The rear (instructor's) cockpit contained a duplicate set of flying and engine controls and all the essential instruments and hydraulic controls. The two cockpits were covered with separate tandem canopies. To provide space for the equipment displaced by the second cockpit the internal armament was reduced to two 20mm cannon. The T Mk 20 was able to carry the normal underwing load of rockets and bombs. To compensate for the change of centre of gravity caused by the introduction of the second cockpit, there was a slight increase in the areas of the tailplane and elevators. A fixed tailwheel was introduced and the aircraft had folding wings and an arrester hook for carrier operations.

The first Sea Fury T Mk 20, VX818, was flown at Langley on 15 January 1948 and by early March the cockpit had been modified by the introduction of a connecting transparent 'tunnel' between the two canopies. An oblique mirror mounted above the gun sight in the rear cockpit provided in effect a periscope that enabled the instructor to take aim over the head of the pupil. The position of the mirror above the fuselage had necessitated a four-strut pyramid support which apparently had little effect on the airstream. The requirement for deck landing was subsequently abandoned and consequently the arrester hook was deleted.

The success of the development trials resulted in a production order for 60 Sea Fury T Mk 20s being placed for the Royal Navy, and the type entered service on 8 February 1950 when the first three production aircraft, VX280, VX281 and VX282, were delivered to No. 1 Aircraft Receipt and Despatch Unit (RDU) at RNAS Anthorn (HMS *Nuthatch*). The first two of these were issued to No. 736 Squadron at RNAS Culdrose (HMS *Seahawk*) and the third to the Station Flight at RNAS Lee-on-Solent (HMS *Daedalus*). The Fleet Air Arm found the T Mk 20 to be invaluable, using it for a variety of tasks including conversion and instrument flying training, communications and, at times, on normal squadron activities to supplement the

Sea Fury T Mk 20, VX308/201/CW, of No. 736 Squadron, Culdrose, 1951. (A. E. Hughes)

Sea Fury FB Mk 11s. In all, Sea Fury T Mk 20s were used by three first line squadrons, nine second line squadrons and five RNVR squadrons, along with several station flights.

Several records were established by the Fury family

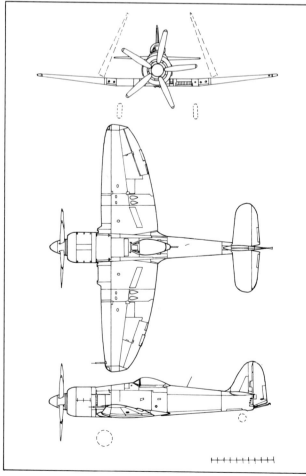

7   Sea Fury FB Mk II

of aircraft. These included the London to Malta record when, on 19 July 1949, a flight of four Fleet Air Arm Sea Fury FB Mk 11s led by Lt. Cdr. W. R. MacWhirter DSC, flew the 1298 miles from London to RNAS Hal Far, Malta, in 3 hours 23 minutes. On 22 July 1950, during the Air League Challenge Cup Race at Sherburn-in-Elmet, Frank Murphy, flying a Sea Fury Mk 20, broke the existing record in the British National Class for piston engined aircraft, achieving 377.8 mph.

During the mid-1950s, the Fleet Air Arm began phasing the Sea Furies out of service and putting them into storage. The Hawker Aircraft Company, with an eye to overseas sales, bought back some 205 Sea Fury FB Mk 11s and 38 T Mk 20s. Not all of these aircraft were intended for resale, as many were used as a source of parts for the aircraft selected for reconditioning. Hawker had some success, and aircraft were sold to the air-forces of Pakistan, Burma and Cuba, and in addition a number of T Mk 20s were sold to West Germany for target towing duties.

Production

N.7/43 SR661, SR666, VB857 (Prototypes).
F Mk X TF895 – TF928, TF940 – TF995.
FB Mk 11 TF956 – TF973, TF985 – TF999, TG113 – TG129, VR918 – VR952, VW224 – VW243, VW541 – VW590, VW612 – VW670, VW691 – VW718, VX608 – VX643, VX650 – VX696, VX707 – VX711, VX724 – VX730, VX748 – VX764, WF590 – WF595, WF610 – WF627, WE673 – WE694, WE708 – WE736, WE785 – WE806, WM472 – WM482, WM487 – WM495, WG564 – WG575, WG590 – WG604, WG621 – WG630, WH581 – WH594, WH612 – WH623, WJ221 – WJ248, WJ276 – WJ292, WJ294 – WJ297, WJ229 – WJ301, WN474 – WN479, WN484 – WN487, WZ627 – WZ656.
T Mk 20 VX818, VX280 – VX292, VX297 – VX310, VZ345 – VZ355, VZ363 – VZ372, WE820 – WE826, WG652 – WG656.

CHAPTER EIGHT
Fairey Firefly

Developed during the early years of the Second World War, the Firefly entered operational service in October 1943 and served with distinction in both Europe and the Far East.

In 1943, a Firefly Mk I, Z1855, was fitted with a two-speed, two-stage Rolls Royce Griffon 61 engine and a larger 'beard' radiator to become the prototype of the Mk III. It was discovered that this radiator had an adverse effect on the aircraft's control and stability, which led to the version being abandoned. The following year, to resolve this problem, another Firefly Mk I, Z2118, had its original radiator replaced by smaller units in the leading edge of each inner wing. As part of the Firefly development programme, Z2118 and three other Mk Is, Z1835, MB649 and PP482, had a 2,300hp, two-speed, two-stage Rolls Royce Griffon 72 engine installed, driving a four bladed propeller. This modification, together with the introduction of wing leading

edge radiators, clipped wings and a dorsal fin, ultimately produced the Firefly Mk IV. On the satisfactory completion of trials, which included operations from HMS *Illustrious* with PP482, a production contract was placed with Fairey for 293 FR Mk IVs, including the conversion of 53 Mk Is already on the production line to FR Mk IV standard. The first production conversion, TW687, made its initial flight on 25 May 1945. Production aircraft differed from the development aircraft only by the introduction of the 2245hp Griffon 74 engine and, in the case of the later aircraft, the deletion of the small air intake under the nose. With the end of the war, 133 Firefly Mk IVs were cancelled, and of the 160 Firefly Mk IVs built, 40 were delivered to the Royal Netherlands Naval Air Service during 1947.

Following service trials by No. 778 Squadron commanded by Lt. Cdr. R. H. P. Carver DSC, at RNAS Ford during 1947, No. 825 Squadron of the Royal Canadian Navy was the first to operate the FR Mk IV when it re-equipped with nine of them at RNAS Eglinton on 8 August 1947 and embarked on HMCS

Firefly FR Mk IVs, TW735 and TW724, during carrier trials on HMS *Illustrious*, March 1947. (RAF Museum)

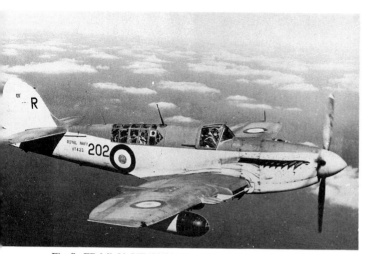

Firefly FR Mk V, VT423/202/R, of No. 812 Squadron, HMS *Glory*.

Magnificent on 22 May 1948 for the voyage to Canada. The first Royal Naval squadron to receive them was No. 810 commanded by Lt. Cdr. L. R. Tivy, which re-formed at RNAS Eglinton with 12 aircraft on 1 October 1947 as part of the 17th Carrier Air Group. The squadron flew its aircraft on board HMS *Implacable* for the first time on 4 May 1948, operating from there for the next 10 days before flying to its new shore base, RNAS *Donibristle* prior to joining HMS *Theseus* and spending some four months at sea on a cruise to South Africa. This was followed by a two month Mediterranean

Firefly FR Mk V, WB421/238/0 of No. 810 Squadron, HMS *Ocean,* at K-3 Air Base in Korea during August 1953. (H. Holmes)

cruise during which the squadron spent some two weeks operating from Malta. There was a final cruise in Home waters during May 1949 and, after taking part in army exercises in Germany in early October 1949, the squadron disbanded at RNAS St. Merryn on the 16th of that month.

Only two other Fleet Air Arm squadrons were equipped with Firefly FR Mk IVs. The first of these was No. 812, commanded by Lt. Cdr. F. G. B. Sheffield DSC, which re-equipped in March 1948 at RNAS Ford, only to have them replaced by Firefly Mk 5s four months later. The second squadron, No. 814, commanded by Lt. Cdr. F. A. Swanton DSO, DSC and Bar, re-equipped in April 1948 at RNAS Lee-on-Solent shortly before moving to RNAS Eglinton on 10 May 1948. While operating FR Mk IVs, the squadron spent four periods on board HMS *Vengeance*, which included a cruise to South Africa and cold weather trials within the Arctic Circle. By February 1949 the squadron had started to take delivery of Mk 5s and the type was completely superseded in November.

The only other first line squadron to be equipped with the Firefly FR Mk IV was No. 816, a Royal Australian Navy squadron which reformed at RNAS Eglinton on 28 August 1948 under the command of a Fleet Air Arm Officer, Lt. Cdr. C. R. J. Coxon. After working up at Eglinton, the squadron embarked on HMAS *Sydney* on 8 February 1949 for the voyage to Australia. From its shore base, RANAS Nowra and from HMAS *Sydney,* the squadron continued to operate FR Mk IVs alongside Mk 5s from March 1949, until the last Mk IV was phased out in October 1951. During

Firefly AS Mk 6s, WJ111/225/J and WJ1116/259/J, of No. 814 Squadron, each armed with 16 rocket projectiles, on HMS *Eagle*. (B. J. Lowe)

April 1951, No. 1840 Royal Naval Volunteer Reserve Squadron commanded by Lt. Cdr. N. H. Bovey DSC, formed as an Anti-Submarine Squadron at RNAS Culham, equipped initially with six FR Mk IVs as an interim arrangement until they were replaced by Firefly AS Mk 6s in July 1951.

This was, however, not the end of the Mk IV in Fleet Air Arm service as a quantity were converted retrospectively into target tugs by the installation of a ML Type G winch under the wing centre-section. The first squadron to receive the TT Mk IV was No. 771, the Southern Fleet Requirement Unit commanded by Lt. Cdr. J. A. Welphy and based at RNAS Lee-on-Solent, which received its first aircraft during November 1951. The headquarters of the squadron was moved to Ford on 1 September 1952, where it continued to operate the TT Mk IVs along with Firefly FR Mk Is and Sea Vampire F Mk 20s until disbanded in August 1955. At this time, No. 771 Squadron's fleet requirements activity was taken over by No. 700 Squadron, commanded by Lt. Cdr. R. W. Turrol, and was also based at Ford. This squadron operated the TT Mk IVs until February 1957, when the last of the type was phased out of service.

Following the Mk IVs down the production line were the Firefly Mk 5s, which externally were identical to the Mk IVs differing only in internal details. Three different versions of the Mk 5 were produced; for fighter reconnaissance, night fighter and the anti-submarine roles designated respectively the FR, NF and AS. The initial production aircraft made its first flight on 12 December 1947. While Mk 5s were still in production a new version was developed, identified as the Firefly AS Mk 6. This aircraft, basically similar to the AS Mk 5, carried no armament but was equipped with the latest

British sonobuoys and equipment. Production of the Mk 6 began with WB505, which flew for the first time on 23 March 1949. A number of changes were introduced during the Mk 5 production run, the most significant being the incorporation of powered wing folding, introduced on VX414, the 240th Mk 5. In all, a total of 352 Firefly Mk 5s and 133 Firefly Mk 6s had been built when production came to an end in 1951. These totals included a number of aircraft for the Royal Netherlands Naval Air Service, the Royal Canadian Navy and the Royal Australian Navy.

Firefly Mk 5s entered squadron service in May 1948 when they joined No. 778 Squadron, a carrier trials unit at RNAS Lee-on-Solent and the high speed flight of No. 782 Squadron at RNAS Donibristle. No. 812 Squadron, commanded by Lt. Cdr. F. B. G. Sheffield, was the first front line squadron to receive the Firefly Mk 5 when it replaced its Mk IVs with twelve Mk 5s in July 1948. The following month the squadron sailed in HMS *Ocean* to the Mediterranean, where it was to spend the next three years sharing time between its shore base, RNAS Hal Far, and the carrier HMS *Ocean,* which was replaced by HMS *Glory* in November 1949. On its arrival at Hal Far in September 1948, No. 812 Squadron expanded to include a flight of four Firefly NF Mk Is. During 1949, four Fleet Air Arm squadrons, Nos. 804, 810, 814 and 816, and additionally one Royal Canadian Navy squadron, No. 825, re-equipped with Mk 5s. These were followed by the Royal Australian Navy Squadron No. 817 in 1950 and No. 820 Royal Navy squadron in 1951. In May 1951 the Royal Canadian Navy Squadron No. 825 was re-numbered No. 880, and on 12 June 1951 No. 825 Squadron commanded by Lt. Cdr. C. K. Roberts reformed as a Fleet Air Arm squadron with eight AS Mk 5s at RNAS Eglinton.

Heading the British fleet off Korea in 1950 was HMS

Firefly T Mk 1, Z2027, 27 November 1947. (RAF Museum)

Triumph carrying No. 800 Squadron equipped with Seafire FR Mk 47s, and No. 827 Squadron equipped with Firefly FR Mk Is. The two squadrons carried out strafing, bombing and rocket attacks on coastal shipping and shore targets, the Fireflies additionally carrying out anti-submarine patrols. Shortages of spare parts and replacement aircraft created serviceability problems which limited the squadron's operations, so that by the time HMS *Triumph* sailed for home in October after being relieved by HMS *Theseus,* there were only three operational aircraft left. However, when HMS *Theseus* started operations the supply problems had been virtually resolved. In six months the 17th Carrier Air Group, comprising the Sea Fury FB Mk 11s of No. 807 Squadron and the Firefly AS Mk 5s of No. 810 Squadron, flew a total of 3446 sorties, many during the appalling weather of the Korean winter of 1950/51. During the remainder of the Korean war, either a British or Australian carrier was on station maintaining a blockade of the west coast. Each of these carriers operated a Carrier Air Group which comprised a squadron of Sea Fury FB 11s and a squadron of Firefly Mk 5s. The other Firefly squadrons to operate off Korea were No. 812 Squadron in HMS *Glory,* No. 825 Squadron in HMS *Ocean,* and No. 817 Squadron in HMAS *Sydney.* In 1953 Nos. 825 Squadron (Firefly FR Mk 5s) and 802 Squadron (Sea Fury FB Mk 11s) were jointly awarded the Boyd Trophy for 1952 for their activities in Korea, which included achieving flying 123 sorties in a single day, firing 6000 rockets and dropping 4000 bombs during their tour of duty aboard HMS *Ocean.*

With the increasing importance of anti-submarine warfare, the more specialised Firefly AS Mk 6, with its improved equipment, began to enter front line squadron service. In December 1951, Royal Australian Navy Squadron No. 817 began to re-equip with the type, and by the time the AS Mk 6 began to be phased out of front line service by the Fleet Air Arm it had equipped seven Fleet Air Arm and two Royal Australian Navy squadrons. In addition, AS Mk 6s had been used by 13 second line squadrons and had equipped the four RNVR anti-submarine squadrons. The RNVR squadrons started to lose their Fireflies in November 1955 when No. 1830 Squadron at RNAS Abbotsinch re-equipped with Grumman Avenger AS Mk 5s, No. 1841 Squadron at RNAS Stretton being similarly re-equipped the following month. Then, in May 1956, No. 1840 Squadron at RNAS Ford changed over to Fairey

Gannet AS Mk 1s, and finally, in June 1956, No. 1844 Squadron at RNAS Bramcote had its Fireflies replaced with Avengers.

Towards the end of the Second World War it was decided that a trainer version of the Firefly would prove invaluable for the type conversion of pilots. Work was immediately put in hand at Fairey's Stockport factory, and a prototype for the Firefly T Mk 1 was made by converting a Firefly FR Mk I, MB750, to have a rear instructor's cockpit at the observer's position similar to the standard pilot's cockpit. This was provided with a full set of controls and instrumentation under a typical Firefly pilot's sliding canopy. The location provided the instructor with an excellent view, almost as good as the pupil's. MB750 flew for the first time from Ringway in July 1946, carrying the company's class 'B' markings 'F 1'. It was later used by the company as a demonstrator and registered G-AHYA. On 31 August 1947, flown by Group Capt. R. G. Slade, it took third place at the Lympne High-Speed Handicap race, with an average speed of 290 mph. In all 34 Firefly T Mk 1s were converted at Stockport, with the first production aircraft MB473 flying at Ringway for the first time on 1 September 1947. Only nine of the T Mk 1s were fitted with armament, and this comprised two wing mounted 20mm cannons.

The Firefly T Mk 2 was virtually identical to the T Mk 1, but was an operational pilot-trainer and consequently all the 57 built were armed with two 20mm cannons in the outer wing and synchronised gyro-gunsights being provided in both cockpits for the pupil and the instructor. It was normal for a single Firefly Trainer to be attached to each of the Firefly equipped RNVR squadrons although, for a short period, two of the other RNVR fighter squadrons also had a Firefly Trainer on their strength.

The last of the trainer versions of the Firefly was the T Mk 3, which was a conversion of the Firefly FR Mk I, virtually identical to the existing FR Mk I but with the armament deleted, ASH radar in a pod mounted under the nose and special equipment installed in the rear cockpit for training observers for the anti-submarine role. The T Mk 3 was the basic equipment of No. 796 Squadron, the Observer's School at RNAS Eglinton, from July 1950 to June 1953. The main disadvantage of this version was that it was only possible to accommodate one trainee observer in the rear cockpit. The Firefly T Mk 3s were also the initial equipment for most of the RNVRs anti-submarine squadrons, being re-

Firefly T Mk 2, MB578/997, of No. 738 Squadron, Culdrose, 1950. (A. E. Hughes)

Sea Vampire T Mk 22, XA107/681, of Lossiemouth Station Flight, 1966. (G. A. Jenks)

favourably with that being achieved by contemporary piston engines. As for the Vampire itself, its very limited endurance ensured that it would never achieve operational status with the Fleet Air Arm. Even so the Admiralty considered that it had a role to play in the introduction of jet powered flight into the Fleet Air Arm. Consequently an order for 30 Sea Vampire F Mk 20s was placed and these were delivered during 1948 and 1949. The F Mk 20 was basically the RAF's Vampire F Mk 3, differing only in the introduction of naval equipment. The small size of the Vampire enabled it to be operated aboard carriers without any requirement of folding wings, thus helping to keep it a relatively simple and therefore economical aeroplane. The majority of the production Sea Vampires were to serve with various second line squadrons, initially for trials and later in a training role. Squadron detachments operated Sea Vampires on board carriers from time to time, providing the Royal Navy with much useful experience of handling jet aircraft at sea. In addition, the aircraft was instrumental in introducing a large number of navy pilots to jet flying.

It was a Sea Vampire F Mk 20 flown by Rear Admiral Couchman that was given the honour of leading the flypast of over 300 Fleet Air Arm aircraft during the Queen's Coronation Review of the Fleet at Lee-on-Solent on 13 June 1953.

The No. 806 Squadron aerobatic display team formed in 1948 to tour the USA and Canada was equipped with Sea Furies, Sea Hornets and a single Sea Vampire F Mk 20, VF315. This aircraft, which was built as a Vampire F Mk 1 before conversion for the Royal Navy, was the first Royal Naval jet to appear in the USA. Flown by the squadron's commanding officer Lt. Cdr. D. B. Law DSC, it was used as the grand finale

bringing each display to a conclusion with a most impressive exhibition of low level, high speed aerobatics.

With the advent of jet-powered aircraft the RAE at Farnborough began a research programme to consider the feasibility of naval aircraft operating without an undercarriage, using a catapult launch for take-off and landing on to a flexible rubber deck. The exclusion of a retractable undercarriage, it was thought, would offer considerable savings in weight and cost. An early production Vampire F Mk 1, TG286, was selected to take part in the trials and was modified to Sea Vampire F Mk 21 standard. For the trials a flexible deck was installed on the airfield at Farnborough during the second half of 1947 and initial trials commenced with the aircraft being dropped vertically on to its underside from various heights to check the ability of the deck to absorb the energy generated without causing damage to the airframe. The actual flying trials commenced on 29 December 1947 with the aircraft, flown by Lt. Brown, coming to grief on the first attempt. The speed on the approach was allowed to drop too low and the engine failed to respond quickly enough to the pilot's urgent demand for more power. This resulted in the aircraft hitting the forward edge of the flexible deck and bouncing off to crash on the grass. Fortunately the pilot was unhurt, but the aircraft was badly damaged. Trials did not re-start until March 1948 with the first landing taking place on 17 March, and by the end of the trials Lt. E. M. Brown had carried out 40 landings on the flexible deck at the RAE.

A flexible deck was installed on the light fleet carrier HMS *Warrior*, and using the second prototype Vampire, LZ551, Lt. Brown successfully made the first landing on a flexible deck at sea on 3 November 1948. Although the sea trials were completed satisfactorily, it became evident that the difficulties of handling aircraft without undercarriages on a carrier deck created operational problems which would be very difficult to resolve and consequently the scheme was abandoned.

In 1950 the DH115, a trainer version of the two-seater DH113 Vampire NF Mk 10 night fighter was developed, the prototype carrying class "B" markings G-5-7 (subsequently WW456), flying for the first time on 15 November 1950. In addition to the prototype, two pre-production aircraft, WW458 and WW461, were built and both of these were delivered to RNAS Culham in 1952 for evaluation by the Fleet Air Arm. On completion of these trials, a production order was placed for 53 aircraft designated Sea Vampire T Mk 22 for the Fleet Air Arm and an additional 20 were ordered later, all being built by de Havilland at its Christchurch factory. The fourth production aircraft, XA103, was the first to be delivered to the Royal Navy, being taken on charge at RNAS Stretton on 18 September 1953. All the Sea Vampire T Mk 22s had been delivered by the end of May 1955.

In addition to their use as trainers by the Nos. 718, 736, 759 and 764 training squadrons, the Sea Vampire

T Mk 22s were also used for instrument rating and communications by various RNVR and first line squadrons and also by several station flights. At least three T Mk 22s, XA154, XA160 and XG775, finished in a navy blue colour scheme, were used as Admiral's Barges by the Flag Officer Flying Trainer (FOFT).

Initially the Sea Vampire T Mk 22s were not equipped with ejection seats but subsequently Martin Baker ejection seats were fitted during a retro-fit programme at the de Havilland factory at Broughton, near Chester during 1956. The Sea Vampire F Mk 20s were withdrawn from service in 1957 and scrapped, while the T Mk 22s started to be withdrawn from service in the mid 1960s when a quantity of them were purchased by Hawker Siddeley Aviation with the intention of refurbishing them and offering them for sale overseas; but this venture did not meet with a great deal of success. A small number were retained by the Fleet Air Arm and used primarily by Station Flights and the FRU until 1970.

Production

F Mk 20 VV136 – VV 165.
T Mk 22 WW458, WW461, XA100 – XA131, XA152 – XA172, XG742 – XG749, XG766 – XG777.

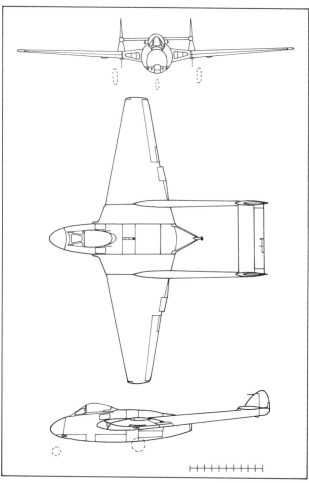

9   Sea Vampire F Mk 20

CHAPTER TEN

Fairey Spearfish

In 1943 the Admiralty issued Specification O.5/43 to cover a projected Barracuda replacement. The design team at Fairey Aviation, led by H. E. Chaplin, started work immediately on the project, to be named Spearfish. The requirement was for a multi-role aircraft primarily for the torpedo/dive-bomber roles, but also suitable for reconnaissance and anti-submarine duties. The large size of the Spearfish was dictated by the requirements to carry a large amount of fuel internally, and to provide a bomb bay large enough to accommodate any of the range of contemporary offensive weapons.

Specification T.21/43 was issued to cover a trainer variant of the Spearfish specification, but it was presumably intended that this was to be developed at some later stage.

A contract was placed on 19 August 1943 for seven prototype/pre-production and 152 production aircraft but, presumably because other projects had higher priorities, progress with the Spearfish was very slow and the first prototype, RA356, did not make its

maiden flight until 5 July 1945 with F. H. Dixon at the controls. The Spearfish was a two-seat aeroplane powered by a single 2500hp 18-cylinder Bristol Centaurus engine driving a Rotol five-bladed propeller. The total fuel capacity of 409 gallons was carried in three wing tanks and this could be supplemented by the installation of a 180 gallon auxiliary tank in the bomb bay. Defensive armament was to comprise two wing mounted 0.5in Browning machine guns, and a remotely-controlled Frazer-Nash FN95 barbette, mounting two 0.5in Browning machine guns installed in the top of the fuselage just aft of the cockpit and controlled by the observer.

The aircraft was of conventional all-metal construction with a light-alloy monocoque fuselage. The wing centre section was built as an integral part of the fuselage and the outer wings were folded hydraulically to be parallel to the fuselage, similar to the Firefly. To provide an adequate take-off and landing performance, very large Youngman type flaps were fitted to the wing trailing edge, inboard of the powered ailerons.

The second Spearfish to fly, RN241, was the first built at Fairey's Heaton Chapel, Stockport, Works and

First prototype Spearfish RA356, 1946. (RAF Museum)

made its first flight at Ringway, now Manchester International Airport, on 29 December 1945. Trials showed that the Spearfish offered little improvement in performance over the Barracuda it was intended to replace. The aircraft did not provide any warning of the approach of the stall which was considered highly desirable for carrier operations. It soon became apparent that both the specification and the resulting aeroplane were not what were required for the Fleet Air Arm. Two further Hayes-built Spearfish, RA360 and RA363, flew during 1947, but by this time the project had been cancelled. The final Spearfish to be built, TJ175, completed at Stockport, was never flown

following the cancellation of the last two prototypes, TJ179 and TJ184, and the production contract.

Following cancellation, the three Spearfish prototypes were used for some time in a variety of experimental roles. RA356 was used by the Naval Aircraft Department at RAE Farnborough for the development of new arrester gear before being transferred to Napier at Luton where it was used for icing trials. The second prototype was taken on charge by the Carrier Trials Unit at RNAS Ford, who used it as the unit's hack until 1952. The third prototype, RA363, crashed on 1 September 1949 during radar trials at the Royal Radar Establishment at Defford.

Production

Prototypes RA356, RA360, RA363, RN241, (TJ175 not flown), (TJ179, TJ184 cancelled).

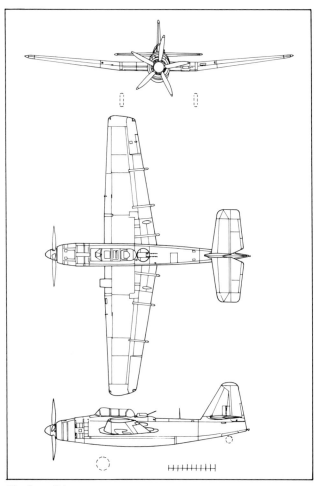

10   Spearfish

CHAPTER ELEVEN
Supermarine Seagull

In an attempt to continue their successful line of small amphibious aircraft beyond the Walrus and Sea Otter, Supermarine started work in 1940 on the design of an aircraft to meet Specification S.12/40 for an advanced fleet reconnaissance amphibious aircraft to operate from aircraft carriers and the larger warships. The initial design work was delayed when the Supermarine works at Woolston in Southampton was bombed and the design team had to be transferred to facilities at Southampton University. The design study was submitted to the Air Ministry in October 1940 and offered both monoplane and biplane versions of the project, to be powered by either the Rolls Royce Merlin 30 or the Bristol Taurus engine. The projected design, identified as the Supermarine Type 347, was relatively small, to be operated by a crew of four, and equipped with Air-to-Surface-Vessel (ASV) radar. The defensive armament consisted of a four-gun turret located behind the engine nacelle.

First prototype Seagull ASR Mk 1, PA143, October 1948.
(RAF Museum)

In the years immediately prior to the Second World War, moves towards monoplanes coupled with increasing landing speeds were making operations on aircraft carriers in particular much more hazardous. Various high lift devices, including slots and flaps, were introduced which, although reducing the landing speed to a more suitable level, resulted in a pronounced nose-up landing approach which restricted forward vision, making landings on a carrier deck particularly difficult. In an attempt to resolve this problem, Supermarine introduced a variable incidence wing on their pre-war Supermarine Type 322 project designed to meet Specification S.24/37. The variable incidence wing allowed the fuselage to remain level as the speed reduced and the angle of incidence of the wing increased. Supermarine, being totally convinced of the soundness of this concept, included a variable incidence wing on the Type 347 project from the outset.

Throughout 1941 and 1942, an intensive research programme on variable incidence wings was carried out jointly by Supermarine and the RAE at Farnborough,

primarily using models in the wind tunnel but also fitting wings with slotted flaps and leading edge slats to a Miles M18 light aircraft.

The programme was much delayed, and it was not until 9 April 1943 that the company received an Instruction to Proceed (ITP) for the design and manufacture of three aircraft to Specification S.12/40, and serial numbers PA143, PA147 and PA152 were allocated.

It was understood from the outset that mounting the engine and wing on a pylon above the fuselage was likely to create problems for the design of the tail surfaces, so further wind tunnel tests were carried out at the RAE early in 1944 and resulted in a high mounted tailplane with a large angle of dihedral. A fin and rudder was mounted on each end of the tailplane. In November 1944 Specification S.12/40 was superseded by S.14/44 for an amphibious Air-Sea Rescue and Reconnaissance aircraft. To meet the new specification the design was changed by the removal of the gun turret and the introduction of the Rolls Royce Griffon engine rather than the Merlin or Taurus. This revised project was identified as the Supermarine Type 381 and named "Seagull".

Although of rather unusual appearance, the Seagull was of conventional stressed skin construction. The variable wing incidence was provided by two electrically operated screw jacks which could alter the incidence up to a maximum of 8.5 degrees. A hydraulically operated undercarriage was provided, with the main undercarriage retracting into the sides of the fuselage and the tailwheel folding backwards to blend into the rear fuselage. For carrier operation an arrester

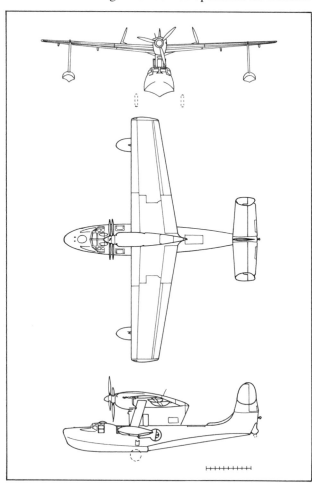

11 Seagull ASR Mk 1

hook was provided. Power was provided by a 1815 hp Rolls Royce Griffon 29 engine driving a Rotol 10ft 4in diameter six-bladed, contra-rotating propeller.

Taxying trials with PA143 commenced on 21 June 1948, and the first flight took place on 14 July 1948 from Southampton Water with Supermarine's chief test pilot M. J. Lithgow at the controls. Following ground handling trials at RNAS Lee-on-Solent the aircraft was flown to Supermarine's airfield at Chilbolton. The first public appearance of PA143 was at the SBAC Flying Display, Farnborough, during September 1948 when it was demonstrated by M. J. Lithgow.

Although initially suffering from buffeting and stability problems, the test flight programme continued steadily with the problems being resolved, generally by minor modifications which included the introduction of a central fin. The Seagull was cleared for production at the Final Conference in May 1949, but no production contract was placed.

The second prototype, PA147, flew for the first time on 2 September 1949 and shortly afterwards was demonstrated by Lithgow at the SBAC Flying Display at Farnborough. Before the end of September PA147 was delivered to the RAE Farnborough for landing trials on the dummy deck. Following the successful completion of these trials, the aircraft was delivered to RNAS Culdrose from where it was flown by Lt. D. G. Parker on 21 October 1949 to the deck of HMS *Illustrious* for carrier trials. By the end of the trials on 27 October a total of 54 landings had been made.

The first Seagull PA143 was criticised by the Marine Aircraft Experimental Establishment (MAEE) at RAF Felixstowe for its poor ground handling and its inability to turn on water in winds of over 15 knots. These handling problems were resolved by increasing the areas of the water rudder and the central fin and rudder, and by introducing smaller outboard fins and rudders.

Supermarine entered PA147 in the Air League Cup Race at Sherburn-in-Elmet on 22 July 1950, and it was flown into fourth place by the company's chief production test pilot L. R. Colquhoun. In this race Colquhoun also established the world's amphibian 100Km closed circuit record at a speed of 241.9 mph.

As time passed the prospects of a production order steadily diminished as it was becoming evident that helicopters would take over the Air-Sea Rescue role from the obsolescent Supermarine Sea Otter. Trials with the first two prototype Seagulls continued, although completion of the third prototype had been abandoned some time earlier with construction at an advanced stage. PA147 was sent to the MAEE for further trials in September 1951, but these were brought to a premature end when a fire on board caused an extinguisher to explode, seriously damaging the fuselage. It was considered that the repairs would be very expensive and, as there was no intention to place a production order for the type, it was decided to cancel the project and all three prototypes were disposed of as scrap in 1952.

Production

ASR Mk 1 (Prototypes) PA143, PA147, PA152.

CHAPTER TWELVE
Westland/Sikorsky Dragonfly

Earlier trials and evaluation of the Hoverfly helicopter by the Royal Navy had demonstrated that, although the Hoverfly itself was unsuitable for operational use, helicopters would be ideal aircraft for naval operations, offering a wide range of possible uses. Unfortunately, late in World War Two, the British aircraft industry was only just starting to become involved in helicopters with Bristol, Cierva, Fairey and Westland all playing a part in the development of a British helicopter industry. Whereas the other companies were busy preparing designs for new projects, the Westland Aircraft Company decided that it would be more sensible to produce American types under licence as that country's helicopter technology was years ahead of the British at that time. Consequently, during 1946 Westland made approaches to Sikorsky, a part of the United Aircraft Corporation, the leading helicopter manufacturer in the USA, and in December 1946 agreement was reached for the Westland Aircraft Company to build the Sikorsky S51 under licence. The S51, which had flown for the first time in February 1946, was basically the next stage of development from the Hoverfly and was aimed primarily at the US civil market. It was a four-seat, general purpose helicopter which could be used for either passenger or freight transport. West-

land's agreement with Sikorsky not only allowed it to manufacture the S51 but also to carry out any development that was considered necessary and to market the aircraft overseas, except in, naturally, the USA and Canada.

Initially Westland purchased six Sikorsky built S51s from the USA and used them for trials and for demonstrations to prospective customers. Before production could start it was necessary for certain modifications to be introduced, enabling the S51 to be built of British materials and to be fitted with British equipment, including the 520hp Alvis Leonides 50 engine in place of the 450hp Pratt and Whitney Wasp Junior.

The first S51 supplied by Sikorsky was registered G-AJHW and, after being re-assembled at Yeovil, it flew for the first time on 18 April 1947. The remaining five were delivered to Westlands during 1947 and by the end of that year all had flown carrying British civil markings. G-AJHW was loaned to the Royal Navy for trials and evaluation during 1948, and while on charge received Royal Navy markings including the serial number WB220. In 1949 trials were carried out aboard HMS *Vengeance,* which confirmed the suitability of the S51 for the air-sea rescue role, as a replacement for the antiquated Supermarine Sea Otter. Initially a contract was placed with Westlands for twelve Westland/Sikorsky WS-51s, later increased to 13, for the Royal Navy and this was followed shortly afterwards by an

Dragonfly HR Mk 3, WP498/901/J, of HMS *Eagle* Ship's Flight, 1954. (RAF Museum)

order for three for the RAF. By this time the type had been named Dragonfly and the Royal Navy and RAF versions were identified as the HR Mk 1 and the HC Mk 2 respectively. These two versions of the Dragonfly were basically the same, except that the HR Mk 1 was equipped for air-sea rescue and the HC Mk 2 for casualty evacuation. The first six production Dragonflies built by Westland were retained by the company for development work and all flew initially with civil markings, the first of which was G-AKTW on 5 October 1948. Two were later to receive military markings: the first of these, WB810, was supplied to the Ministry of Supply for trials and the second, WZ749, was delivered to the RAF. Two others were subsequently sold to civilian operators.

The Royal Navy Dragonfly HR Mk 1s started to come off the production line in mid 1949 and No. 705 Squadron, the Royal Navy's helicopter training and special trials unit, commanded by Lt. Cdr. S. H. Suthers at RNAS Gosport, received its first Dragonfly on 13 January 1950, starting a programme of replacement for the squadron's Hoverflies. The squadron had received its full complement of Dragonflies by June 1950, with the last Hoverfly being phased out of service the following November. By the end of the year the squadron's main task was to train helicopter pilots for the search and rescue (SAR) role in readiness for the time when the SAR units replaced their Sea Otters with Dragonflies. The Dragonfly's intended primary SAR task was to fly as 'plane guard' in support of operational fixed wing flying from aircraft carriers.

With the exception of aircraft allocated for trials, all the Dragonfly HR Mk 1s were delivered to No. 705 Squadron, which was to operate them until March 1953. However, following the success of the trials, further contracts were placed for an additional 64 Dragonflies for the Royal Navy. The Dragonfly HR Mk 1 had been fitted with a three-bladed rotor; each blade having a tubular metal spar, a wooden leading edge, and had the aft of the spar covered in fabric. This was considered unsatisfactory for operational use and consequently, for the next production batch, new all-metal rotor blades were introduced. At the same time hydraulic servo controls were also introduced. This new version of the Dragonfly was identified as the HR Mk 3

Dragonfly HR Mk 3, VX600/531, of No. 705 Squadron, Culdrose, over HMS *Centaur,* May 1959. (RAF Museum)

and a total of 58 were built for the Fleet Air Arm, the last six being cancelled. Although hardly ideal for the SAR role with its limited cabin capacity and poor performance giving a cruising speed of 85 mph and a range of 248 miles, it was still an improvement on the previous fixed wing SAR aircraft and was the best helicopter available at the time. In 1957 the final version of the Dragonfly, the HR Mk 5, was introduced. No new HR Mk 5s were built, but it is understood that 25 of the earlier marks were converted at the Royal Naval Air Yard at Donibristle by the introduction of a more powerful version of the Leonides engine and a general updating of equipment.

In January 1951 the first Fleet Air Arm SAR unit to be equipped with Dragonfly HR Mk 3s joined HMS *Indomitable.* At about the same time, trials were being carried out with a Dragonfly operating from a small platform fitted to the Royal Fleet Auxiliary (RFA) Fort Duequesne. These trials, which were carried out in a wide range of weather conditions, were a complete success and led to the eventual introduction of helicopter platforms on a large number of Royal Navy and RFA ships.

When the Korean War started in June 1950, helicopters had yet to take over the SAR role in the Fleet Air Arm's carriers and the first Royal Navy carrier on station off Korea, HMS *Triumph,* was still operating a Sea Otter in the SAR role, as was HMS *Theseus* which replaced *Triumph.* However, during operations, a US Navy Dragonfly with its crew was loaned to HMS *Theseus* for 'plane guard' duties and demonstrated its effectiveness by rescuing five aircrew from the sea. Whereas HMS *Indomitable,* the first British carrier to operate Dragonflies in the SAR role, remained in Home waters, the second, HMS *Glory,* was to sail to the Far East, there to relieve HMS *Theseus* in Korean waters. Again the Dragonfly demonstrated its worth by rescuing four pilots from the sea and a further four from behind enemy lines. Subsequently, a total of 14 Royal Navy aircraft carriers and ten Royal Naval Air Stations operated Dragonflies in their SAR flights.

In June 1953 twelve Dragonflies took part in the flypast for the Royal Naval Coronation Review at Spithead. A helicopter flight equipped with Hiller HT Mk 1s was introduced on the Survey Ship HMS *Vidal* in May 1954, but these proved rather unsatisfactory and were replaced by Dragonfly HR Mk 5s, one of which was used to place a British flag on Rockall to confirm British ownership of the rock. Two Dragonflies operating from HMS *Campania* were also involved in providing support for the British nuclear tests at Monte Bello, collecting samples from the area of the explosion for analysis.

In addition to their use for training with No. 705 Squadron and SAR duties with the ship and station flights, the Dragonflies were used for a variety of roles by other units. No. 705 Squadron was the only one to use all three marks of Dragonfly, starting with HR Mk 1s in January 1950 and ending with the phasing out of the last HR Mk 5 in March 1962. When storms in the North Sea seriously damaged the dykes along the Dutch coast in February 1955, there was extensive flooding and the Dragonflies of No. 705 Squadron were involved in the flood relief work, particularly in rescuing people from the roofs of houses. For the work carried out in Holland the Squadron Commander, Lt. Cdr. H. R. Spedding, was awarded the MBE.

Dragonfly HR Mk 5, WG708/960, of the Hal Far SAR Flight 1960, fitted with a 'Sproule net' used to scoop aircrew out of the sea.
(A. E. Hughes)

Among other squadrons that operated Dragonflies was No. 700, which operated a single Dragonfly HR Mk 1, VX597, in support of trials between February and October 1959. No. 701 Squadron, the Helicopter Fleet Requirements Unit at Lee-on-Solent, operated HR Mk 5s from November 1957 to September 1958, while the Fleet Requirements Unit at Hal Far, No. 728 Squadron, operated a number of HR Mk 3s primarily in the SAR role. The Dragonfly HR Mk 3s on the strength of No. 744 Squadron at Eglinton from December 1952 to February 1954 were used for the squadron's secondary duties, that of the Station Flight, and not for its primary role as Naval Air Anti-Submarine Trials and Development Squadron. The only other Fleet Air Arm operators of the Dragonfly were No. 771 Squadron, the Helicopter Trials and Training Squadron at Portland, and finally the Air Experience Flight of No. 727 Squadron at Brawdy, which operated two HR Mk 1s giving helicopter experience to non-flying junior officers from the Britannia Royal Naval College at Dartmouth. The Air Experience Flight was transferred to Roborough in June 1960 and became known as the Britannia Flight. This proved to be the Fleet Air Arm's last operator of the Dragonfly, phasing its last one out of service in June 1967.

During 1954 Westland carried out a major re-design of the Dragonfly which introduced an uprated Leonides engine, made major changes to the fuselage shape to resolve a centre of gravity problem and increased the seating capacity to five. A clamshell door was introduced on the port side of the nose and a new rotor, similar to that used on the WS-55 Whirlwind, was adopted. The new project, named the Widgeon, was carried out as a private venture and the first prototype, G-AKTW, was a conversion of the first Westland Dragonfly. The Royal Navy was impressed by the design and decided that 24 of their Dragonflies should be converted to Widgeon standard and then be identified as Dragonfly Mk 7s. Unfortunately none of these conversions was ever carried out as a shortage of funds in the defence budget led to the cancellation of the programme.

12   Dragonfly HR Mk 3

Production

Prototype (Sikorsky built) WB220.
HR Mk 1 VX595 – VX600, VZ960 – VZ966.
HR Mk 3 WG661 – WG672, WG705 – WG709, WG718 – WG726, WG748 – WG754, WH989 – WH992, WN492 – WN500, WP493 – WP504.

CHAPTER THIRTEEN
Short Sturgeon

The Ministry of Aircraft Production (MAP) issued Specification S.6/43 during 1943 for a high-performance torpedo-bomber/reconnaissance aircraft for use on the large fleet carriers being planned at the time. Three companies, Short Brothers, Armstrong Whitworth and Westland, prepared design studies which were submitted in April 1943. The requirement to include a bomb bay which could accommodate any of the current standard airborne torpedos or up to a maximum of six 500lb bombs and still provide a high performance, prompted the decision to go for a large, by Fleet Air Arm standards, aircraft powered by two engines.

Shortly after the submission of the tenders there was a change in the requirements, with the torpedo role being deleted and a new Specification, S.11/43, issued. The new requirements enabled the fuselage length to be reduced, as the bomb load could be contained within a smaller bomb bay, and consequently the wingspan could also be reduced, making it possible to achieve a higher performance. The revised design studies were submitted to the MAP in June 1943, and in October Short Brothers received a contract for three prototypes of their S38 project. The first two S38 prototypes, RK787 and RK791, were built concurrently by Short's factory at Rochester in Kent. RK787 made its first flight at Rochester Airport on 7 June 1946 with

Geoffrey Tyson at the controls and, by that time, the type had been named the Sturgeon.

Following the surrender of Japan in August 1945, the British Government suspended the construction of the new aircraft carriers and, consequently, the requirement for the Sturgeon FR Mk 1 reconnaissance bomber was cancelled. Fortunately for Short's, it was realised that a target towing aircraft was required for gunnery training, and early in 1946 Specification Q.1/46 was issued to cover a variant of the Sturgeon modified into a fast, carrier-based target towing aircraft. Project S39 was issued to cover the new aircraft, but shortly afterwards Short's adopted the SBAC designation system and the S38 became the SA 1 and the S39 became the SA 2.

Short Brothers started to run down the plant at Rochester, in 1946, reducing it to a service department, and transferred production to Belfast. Although RK787 was flown at Rochester, the incomplete second and third prototypes, RK791 and RK794, were transferred to Belfast for completion. RK791 was completed as the second SA 1 Sturgeon FR Mk 1 prototype, but the third was modified to become the first SA 2 Sturgeon TT Mk 2 prototype and the original serial number RK794 allocated to the aircraft was replaced by the new serial number VR363. Shortly before the transfer to Belfast a fourth prototype, VR371, the second Sturgeon TT Mk 2, was ordered,

Second prototype Sturgeon FR Mk 1, RK791, taking off from HMS *Implacable* during carrier trials in June 1948. (RAF Museum)

Sturgeon TT Mk 2, TS492/594/HF, of No. 728 Squadron, Hal Far, with Mount Etna in the background. (B. J. Lowe)

and this aircraft was built entirely at Short's Queen's Island factory in Belfast.

The Sturgeon was a mid-wing monoplane of conventional light alloy construction, powered by two 2080hp Rolls Royce Merlin 140 engines driving Rotol six-bladed contra-rotating propellers. Armament of the Sturgeon FR Mk 1 was two 0.50in machine guns in the nose, up to eight 60lb rocket projectiles under each wing, and either one 1000lb bomb or two 500lb bombs or four depth charges carried in the fuselage weapons bay. The crew of three comprised the pilot located in a cockpit placed above the wing leading edge and the navigator and radio operator in a cabin located aft of the wing. Being designed for carrier operation, the Sturgeon was equipped with hydraulically operated, folding outer wings, tail mounted arrester hook and catapult points. The Sturgeon TT Mk 2 was basically similar to the FR Mk 1, except for the removal of the armament, the reduction of the crew to two and the introduction of a long nose for a Vinten synchronised camera installation.

Although trials with the prototype Sturgeon FR Mk 1s, including carrier trials on HMS *Illustrious,* were completely satisfactory, no production order was placed and it was the Sturgeon TT Mk 2 that entered production, a contract being placed with Short's for 24 in 1949. Trials using the prototype Sturgeon TT Mk 2 VR363 were carried out during the spring and summer of 1949, including carrier trials on HMS *Illustrious* during May. Following the trials, the elevator area was increased by 20% to improve handling. Further evaluation at the A&AEE Boscombe Down confirmed that the modification was satisfactory and the Sturgeon TT Mk 2 was cleared for service in February 1950.

The Sturgeon TT Mk 2 first entered squadron service in September 1950, when the first aircraft was received by No. 771 Squadron commanded by Lt. Cdr. J. G. Baldwin DSC, at Lee-on-Solent. The squadron operated the Sturgeon alongside Sea Hornets, Sea Mosquitos, Fireflies and Meteors on fleet requirement duties.

The squadron moved to RNAS Ford in September 1952 where it became the Southern Fleet Requirements Unit and two months later its Sturgeons were phased out of service.

The main operator of the Sturgeon was No. 728 Squadron at Hal Far in Malta, which started to receive Sturgeon TT Mk 2s in August 1951 as replacements for its Mosquito TT Mk 39s. All the Mosquitos had been replaced by May 1952. Although the squadron's main task was to provide targets for the Royal Navy's Mediterranean Fleet, they were also used by the British Army and RAF Regiment units on Malta.

In 1951, with the introduction of more advanced radar gunlaying for anti-aircraft guns, the Sturgeon TT Mk 2s were becoming obsolescent. To meet the change in requirement Short Brothers designed a new variant, the SB9 Sturgeon TT Mk 3. The main changes were the removal of the nose camera station and its replacement by a more orthodox shorter nose, and the removal of all equipment that had been required for carrier operations. The first TT Mk 3 was produced by the conversion of Sturgeon TT Mk 2, TS475, at Belfast. A further four TT Mk 2s were converted to Mk 3 standard at the Short Brothers Service Department at Rochester.

The Sturgeon TT Mk 3s entered service alongside the TT Mk 2s of No. 728 Squadron in July 1954 and remained in use until October 1958, almost three years

First prototype Sturgeon TT Mk 3, TS475. (Shorts)

Prototype SB 3, WF632, an anti-submarine aircraft. (Shorts)

after the last Sturgeon TT Mk 2 had been withdrawn.

In 1949, an attempt was made to produce a variant of the Sturgeon to meet Specification GR.17/45 for an anti-submarine search and patrol aircraft. Specification M.6/49 was issued to cover Short's project, identified as the SB3. To meet the requirement, a major redesign of the Sturgeon was carried out. This primarily affected the nose, which was modified to introduce a cabin forward of the pilot's cockpit to accommodate two radar operators and a large radome mounted under the nose. Power was supplied by two 1475hp Armstrong Siddeley Mamba AS Ma3 turbo-prop engines driving four-bladed propellers. Two prototypes of the SB3 were ordered and the first, WF632, was flown for the first time on 12 August 1950 at Belfast by Tom Brooke-Smith, who also demonstrated it at the SBAC show at Farnborough the following month. Unfortunately WF632 proved to be unstable and extremely difficult to trim when flying on one engine. Presumably because of Short's heavy commitments and the potential success of the other projects to meet GR.17/45, the SB3 project was cancelled without any apparent effort to resolve the stability problems and before the second prototype WF636 had flown. Both prototypes were scrapped early in 1951.

Production

FR Mk 1 (Prototypes) RK787, RK791, (RK794 cancelled).
TT Mk 2 VR363, VR371 (Prototypes).
TS475 – TS498.
TT Mk 3 TS475 (Prototypes).
TT Mk 3 TS479 – TS482.
SB3 (Prototypes) WF632, (WF636 not completed).

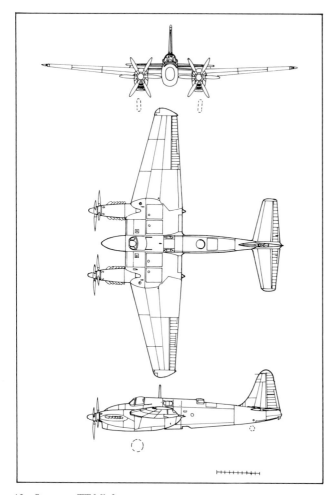

13   Sturgeon TT Mk 3

CHAPTER FOURTEEN

Supermarine Attacker

With the issue in 1944 of Specification E.10/44 for an experimental jet fighter, the Supermarine design team, under the leadership of Joe Smith, produced a design that used the existing Spiteful laminar flow wing. By using the existing wing and undercarriage, the cost and timescale of the development was reduced by a significant amount. Inevitably the wings did require some minor changes; the underwing radiators not required for this project were deleted. Additional fuel tanks were introduced to provide for the increased fuel consumption of the jet engine and to maintain a good landing and take-off performance the flap area was increased. The 4500lb thrust Rolls Royce Nene centrifugal flow engine was selected to power the project. The company received an Instruction to Proceed on 5 August 1944 for the manufacture of three prototypes.

Early in 1945, before manufacture of the prototypes had progressed too far, Specification E.1/45 was issued covering a navalised version of the E.10/44. As a consequence it was decided that only the first prototype should be built to meet E.10/44 and the other two to be built to E.1/45.

A contract was placed on 21 November 1945 for the manufacture of a pre-production batch of six aircraft to E.10/44, by then identified by Supermarine as the Type 392, and 18 aircraft to E.1/45 identified as the Type 398. The first four of the Type 398 aircraft were to have full naval equipment, except for folding wings which were to be introduced on the remaining 14. These hydraulically folding outer wing panels were identical to those used on the Seafang. However, the contract for the pre-production aircraft was cancelled before work on these aircraft had started.

Cancellation of the pre-production batch of aircraft

First prototype Attacker, TS409, September 1946. (RAF Museum)

Attacker F Mk 1s, including WA496/101 and WA494/105, of No. 800 Squadron in 1951. (RAF Museum)

did not affect progress of the prototypes. However, a series of relatively minor modifications to the first prototype, TS409, did delay its completion. TS409 flew for the first time at the A&AEE Boscombe Down on 27 July 1946 with the company's chief test pilot, Jeffrey Quill, at the controls. At this stage the aircraft had an overall polished metal finish. The cockpit was unpressurised and the pilot was not provided with an ejection seat. It was during the early trials that it became apparent that the company's decision to use a tailwheel as opposed to a tricycle undercarriage was to create an unexpected problem and that was that the jet exhaust tended to scorch grass and tarmac. Although the problem was never completely resolved, it was improved by slightly angling upwards the end of the jet pipe.

The second prototype, TS413, flew for the first time at Chilbolton, in north-west Hampshire, on 17 June 1947 in the hands of the company's new chief test pilot Lt. Cdr. M. J. Lithgow. By this time the aircraft had been officially named "Attacker". This was the first naval prototype and except for the lack of folding wings it had all the equipment necessary for operation from a carrier deck. A long stroke undercarriage was introduced, catapult attachment points and an 'A' frame arrester hook were fitted. In addition to the specific naval changes there were other general improvements which included the reduction in size of the fin and the enlargement of the tailplane. The flaps were modified and spoilers were introduced on the upper surface of the wing and the pilot was provided with a Martin-Baker Mk 1 ejection seat.

On 15 October 1947 Lithgow flew TS413 to the deck of HMS Illustrious for the commencement of deck landing trials. During these initial trials, 12 landings were made by three pilots, Lithgow from Supermarine, Lt. E. M. Brown from the RAE, and Lt. S. G. Orr from the A&AEE.

Lt. Cdr. Lithgow flying TS409 captured the 100Km closed circuit world speed record on 26 February 1948 at an average speed of 560.63 mph. The following day he successfully raised the record to 564.88 mph.

The Attacker development programme progressed rather slowly, with the Admiralty apparently unable to make a decision on whether they actually had a requirement for it. Unfortunately the programme hit a further snag on 22 June 1948. The second prototype TS413 was destroyed when it dived into the ground at the village of Bulford during measured take-off and handling trials at the A&AEE, killing the pilot. It was about this time that the Royal Navy started to show renewed interest in the project as a means of the service gaining jet operational experience. Consequently work was put in hand with some urgency to bring TS409 up to Naval Type 398 standard and also to complete the third prototype.

The navalised TS409 rejoined the test flight programme on 5 March 1949 and in November 1949 the Admiralty placed a contract with Supermarine for 60 Attackers, and shortly afterwards a further contract was placed for 36 land based Attackers Type 538 for the Pakistan Air Force. The third prototype, TS416, was finally completed towards the end of 1949 and flew for the first time on 24 January 1950. As with the previous aircraft, TS416 also lacked folding wings, but differed in having the wings 13.5 inches further aft, slightly larger air intakes and an increase in internal fuel capacity. During February 1950 the two surviving prototypes successfully completed a series of 33 landings on board HMS Illustrious.

Attacker production at Supermarine's South Marston factory progressed rapidly and the first production Attacker F Mk 1, WA469, flew for the first time on 5 April 1950. Powered by a 5100lb thrust Nene 3 engine, WA469 was the first Attacker with folding wings and attachment points on the fuselage above the wing trailing edge for the rocket assisted take-off gear (RATOG). Provision was made for the carriage of a 250 gallon ventral fuel tank which was to become virtually a standard fit on all operational aircraft. During one of the early test flights, Supermarine test pilot Leslie R. Colquhoun encountered a serious problem when the starboard wing fold mechanism failed and the outer wing panel lifted into the vertical position, jamming the ailerons. Rather than eject and lose the aircraft, the pilot demonstrated considerable skill in successfully landing the aircraft back at the South Marston airfield, a feat for which he was awarded the George Medal.

As production aircraft came off the track, trials activity intensified and during November 1950 the third and fourth production aircraft, WA470 and WA472 respectively, carried out intensive deck landing trials on HMS Illustrious. In ten days of flying, 197 landings were made. Early production Attacker F Mk 1s were used by Supermarine and the A&AEE for trials. January 1951 was to see Attackers entering Royal Navy service when they were delivered to No. 787 Squadron, the Naval Air Fighting Development Unit of the Central Fighter Establishment, at RAF West Raynham. After only a few had been delivered the fin area was increased by the introduction of a dorsal fin which improved directional stability, this modification being subsequently embodied retrospectively on the earlier aircraft.

In 1951 Supermarine started work on a fighter bomber variant. This resulted in an interim version identified as the FB Mk 1, capable of carrying a single 1000lb bomb or alternatively four 60lb rocket projectiles under each wing. The first aircraft, WA527, flew on 7 January 1952 and the remaining eight aircraft on the production line were completed as FB Mk 1s. In addition seven F Mk 1s were converted to FB Mk 1 standard. A follow-on contract for a further 84 Attackers was placed in 1951, and all these were completed as FB Mk 2s. This version, powered by the Nene 102 engine fitted with an electric high energy starter, also had an accelerator control unit linked to the throttle. A new metal framed cockpit canopy and modified ailerons were also introduced. The first Attacker FB Mk 2, WK319, flew on 25 April 1952, and by the end of 1953 a total of 86 Attacker FB Mk 2s had been delivered, which included two conversions of F Mk 1s, WA491 and WA507.

No. 800 became the Royal Navy's first operational jet squadron when it reformed on 17 August 1951 at RNAS Ford equipped with eight Attacker F Mk 1s. On 4 March 1952 No. 800 took its Attackers aboard HMS *Eagle* for a short period of operational training. However, at this time it had just started to replace the Attacker F Mk 1s with FB Mk 1s and so changing the squadron's role from interceptor to ground attack/ strike. Following shortly after No. 800 was No. 803, which reformed at RNAS Ford on 26 November 1951 equipped with eight Attacker F Mk 1s. Both squadrons formed part of HMS *Eagle*'s 13th Carrier Air Group serving in Home waters during 1953 and in the Mediterranean during the first half of 1954. By the end of 1953 both squadrons had been re-equipped with Attacker FB Mk 2s.

The third operational squadron to be equipped with the Attacker was No. 890 commissioned with eight Attacker F Mk 1s at RNAS Ford on 22 April 1952. It was disbanded on board HMS *Eagle* on 3 December 1952 when its aircraft were transferred to the other two squadrons. During its short existence as an Attacker squadron, 890 had operated all three versions of the aircraft. By the end of 1954 the Attacker had passed out of front line service replaced by the Hawker Sea Hawk.

Attacker F Mk 1s, including WA518/114/J, of No. 803 Squadron and Firebrand TF Mk 5s of No. 827 Squadron on HMS *Eagle* in June 1952. (RAF Museum)

The Attacker had proved to be something of a disappointment in service, its performance above 30000ft being very poor and its rate of climb, particularly for the interceptor role, totally inadequate. When compared with contemporary aircraft with tricycle undercarriages, the forward vision, especially when taxying, was far from good. During high speed combat manoeuvres the wings tended to flex, putting the wing-mounted guns out of alignment, and as a consequence the Attacker proved to be a totally inadequate gun platform.

As part of the programme to re-equip the Royal Naval Volunteer Reserve with jet fighters, No. 718 Squadron reformed at RNAS Stretton on 25 April 1955. Equipped with Attackers and Sea Vampire T Mk 22s, the squadron was soon at work converting the pilots of No. 1831 Squadron of the Northern Air Division from the Sea Fury to the Attacker. No. 1831 Squadron, commanded by Lt. Cdr. P. L. V. Rougier,

Attacker FB Mk 2, WZ294/176/ST, bearing the 'winged greyhound' emblem of No. 1831 RNVR Squadron, Stretton, 1955.

officially reformed as an Attacker squadron on 4 July 1955, making it the first RNVR squadron to be equipped with jet powered aircraft. Almost immediately the squadron formed an aerobatic display team which performed at many air displays during 1955 and 1956, and was probably the only display team to use the Attacker. Nos. 1832 and 1833 Squadrons also re-equipped with Attackers before the end of 1955. Early in 1956 Sea Hawks began to replace the RNVR's Attackers and before the end of 1956, No. 1832 had totally replaced its Attackers. However, conversion of the other two RNVR Attacker squadrons had not started when the Air Branch of the RNVR was disbanded on 10 March 1957

To provide training support for the first line Attacker squadrons, No. 736 Squadron reformed at RNAS Culdrose on 26 August 1952 as the Advanced Jet Flying School equipped with Attackers and Meteor T Mk 7s. On 9 November 1953 the squadron moved to RNAS Lossiemouth where, with No. 738 the piston-engined equivalent to 736 Squadron, it became part of the Naval Air Fighter School. The squadron was by far the largest user of Attackers, equipped with 28 aircraft in May 1954. Two months later the squadron's Attackers started to be replaced by Sea Hawks, and by the end of August the re-equipment was complete. In addition No. 767 Squadron, the Deck Landing Officer Training Squadron based at RNAS Stretton, was equipped with a small number of Attackers for a year from February 1953. The training comprised Airfield Dummy Deck Landings (ADDLs) at Stretton, with periods at sea on either HMS *Triumph* or HMS *Illustrious*. The squadron

was re-equipped with Sea Hawks and Avengers early in 1954.

The Fleet Requirements Unit (FRU) was equipped with Attackers from 1955 until early in 1957. This unit, managed by Airwork Ltd., was operated from the civil airfield at Hurn, RNAS Brawdy and at Brawdy's satellite airfield St. David's, providing targets for the fleet to practice gunnery and radar tracking.

By mid 1957 all the Attackers had been withdrawn from use and stored at the Aircraft Holding Unit (AHU) at RNAS Lossiemouth until they were scrapped in 1958.

The RAF soon lost interest in the Attacker project when it was seen to offer no improvement in performance over their existing Meteors and Vampires. However, it was the start of a line of development that was to lead to the Type 545 Swift which entered RAF service in 1954. Even though the Royal Navy had little interest in these developments, the experimental Type 510, which was in effect a swept wing Attacker, was considered for development into a naval fighter.

The first of the two Type 510s built, VV106, was modified during 1950 by the introduction of an arrester hook and pick-up points for rocket assisted take-off gear (RATOG). Following ADDLs at the RAE Farnborough in September 1950, VV160 was landed on HMS *Illustrious* on 8 November 1950 by Lt. J. Elliot RN and became the first swept wing aircraft to operate from an aircraft carrier. During the trials some 15 landings and take-offs were carried out, including several take-offs using RATOG. Although the trials were considered to be satisfactory, the naval equipment was removed shortly afterwards and no further naval involvement in the programme was to take place.

During 1950 the Attacker was used as the basis of a high speed research aircraft for use in the development programme for the Handley Page HP80 Victor bomber. For this purpose the fuselage of an Attacker was fitted with a crescent wing and a high mounted tailplane. The project, designated the HP88, was designed by Handley Page but constructed by Blackburn and General Aircraft Ltd. at Brough. The aircraft, serial number VX330, was flown for the first time at Brough on 21 June 1951 by Blackburn's chief test pilot G. R. I. 'Sailor' Parker. After a total of some 30 flights VX330 was destroyed on 26 August 1951 when it disintegrated during a low level demonstration flight at Stansted, killing Handley Page's deputy chief test pilot D. J. P. Broomfield.

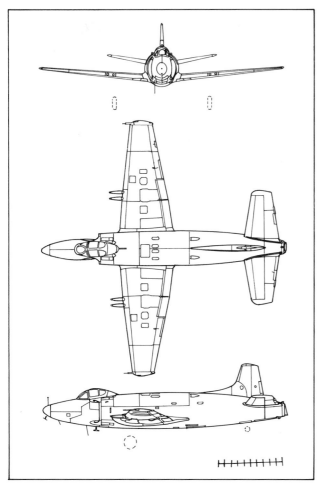

14   Attacker F Mk 1

Production

Prototypes TS409, TS413, TS416.
F Mk 1 WA469 – WA498, WA505 – WA526.
FB Mk 1 WA527 – WA534, WT851.
FB Mk 2 WK319 – WK342, WP275 – WP303, WZ273 – WZ302.

CHAPTER FIFTEEN
Douglas Skyraider

In 1943 the Douglas Aircraft Company started work on a US Navy requirement for a carrier borne bomber to replace the SBD Dauntless. The project, identified as the XSB2D-1, was a single-engined, low wing monoplane with a tricycle-undercarriage and operated by a crew of two. Towards the end of 1943, after a contract had been placed for two prototypes, the requirement changed from a bombing to a torpedo role, identified as BTD-1, which resulted in the deletion of the second crew member. The two prototype XSB2D-1s were built and these were followed by 28 BTD-1s, the aircraft now being named the Destroyer. But in 1944 the US Navy cancelled the project, having changed the requirement to one for an attack aircraft. Douglas immediately set to work and within a few days had prepared a design, the XBT2D-1, to meet the new requirement. This differed from the Destroyer primarily by the introduction of a tail wheel undercarriage and some aerodynamic refinements to improve the aircraft's load carrying ability. The engine selected for the project was the 2500hp Wright Cyclone R-3350-24W.

In all a total of 25 XBT2D-1s, called Destroyer IIs, were built, and the first flew on 18 March 1945. In February, the name was changed to Skyraider, and in the following April, when 20 aircraft had been delivered, the US Navy's designation system had been changed and, to meet with the new system, the XBT2D-1 became the XAD-1. By mid 1945 contracts had been placed for 578 AD-1 Skyraiders, but with the end of the war in August 1945 the quantity was initially reduced to 377 and subsequently to 277.

At this time the low level attacks on the allied Pacific Fleet by Japanese Kamikaze aircraft were causing considerable concern, as it was evident that these aircraft were not being detected early enough to give the fleet adequate time to prepare its defences. In response to this problem, the British issued a requirement in 1943 for an airborne radar system. Unfortunately at the time resources were already overstretched, so no progress was made in developing this type of radar. However, in 1944 a joint Anglo-US Scientific committee considered the problem and decided that the design, development and manufacture of the radar should be carried out in the USA. The requirement was for a radar that could provide early warning of surface vessels and low flying aircraft. In the USA the project was given to the

Massachusetts Institute of Technology at Mount Cadillac, where it was appropriately known as the 'Cadillac' project. In February 1945 a Grumman TBM Avenger, the first aircraft fitted with the new airborne early warning (AEW) radar, was flying, and even with development at a very early stage, work was put in hand to install the new radar in 35 Avengers, which were immediately dispatched to the Pacific War Zone. However, none of the AEW Avengers had actually achieved operational status with the Pacific Fleet before the end of hostilities.

The Skyraider was also considered for use as a vehicle for the AEW system and of the batch of 25 development XBT2D-1 (XAD-1) Skyraiders, the 23rd was converted into the prototype AEW version and identified as the XAD-1W. The 25th was converted into the prototype of a radar countermeasures version and identified as the XAD-1Q. The AEW Skyraider looked very different to the standard AD version primarily because of the very large ventral 'Guppy' radome, which necessitated the fitting of large fixed leading edge slats to the outer wings and small auxiliary fins to the tailplane.

Presumably because of problems with the radar system in the AEW Avengers that had entered service in 1945, no AEW versions of the Skyraiders were built on the initial production contract. In service, the AD-1 Skyraiders suffered a series of structural failures to the main undercarriage and wing centre section. These problems were resolved with the next mark, the AD-2, which included modifications to the undercarriage and the introduction of reinforcing to the wing centre-section structure. In addition, a new cockpit layout, including a new pilot's seat was introduced, although the most significant change was the introduction fo the 2700hp Wright Cyclone R-3350-26WA engine. Of the 356 AD-2s initially ordered only 178 were built – 156 AD-2s and 22 AD-2Qs – but still no production AEWs.

The next mark, the AD-3, included a redesigned cockpit and further structural reinforcing. Included in the 194 AD-3s built were 31 of the AEW version, the AD-3W. In service, the AD-3 demonstrated that the company had finally resolved the structural problems and, from this version onwards, the Skyraider earned a reputation for being tough, and able to absorb a significant amount of battle damage.

Skyraider AEW Mk 1, WT944/301/CW, of No. 778 Squadron, Culdrose, December 1951.

The AD-4 saw the introduction of the uprated Wright R-3350-26WA engine and, in the AD-4W variant, the introduction of the vastly improved APS-19A radar. Being the version current at the time of the Korean war, large orders were placed for the AD-4 totalling 1,051 aircraft, which included 168 AD-4Ws. Production of the Skyraider continued after the Korean war with the AD-5, which introduced side-by-side seating for the pilot and co-pilot/radar operator. In addition to the versions of the AD-5 to undertake the normal specialised roles of attack, AEW, and radar countermeasures, the Skyraider could also be used as a military passenger transport, capable of carrying eight fully armed troops or four stretcher cases in the rear fuselage. A total of 670 AD-5s were built including the 218 AD-5Ws, which made it the most numerous of the AEW versions of the Skyraider. The final two versions of the Skyraider, the AD-6 and AD-7, which were produced specifically for the day attack role. By the time production of the Skyraider came to an end in 1957, a total of 3180 had been built, all for the US Navy. Subsequently, quantities of ex-US Navy Skyraiders were supplied to various countries. Attack versions were supplied to France, Cambodia and Vietnam, and were to remain in service in South-East Asia until the late 1970s.

It was Britain, however, which was the first country to receive Skyraiders when in 1951 50 AD-4Ws, the AEW variant, were supplied to the Royal Navy under the Mutual Defence Aid Programme (MDAP). These aircraft were identified as the Skyraider AEW Mk 1 in Fleet Air Arm service and of the 50 aircraft supplied, 20 were new build diverted on the production line at the Douglas Aircraft Company's plant at El Segundo, and 30 were refurbished ex US Navy aircraft. Deliveries commenced in November 1951 when the first four Skyraiders were delivered by sea in the SS *American Clipper,* arriving at the docks in Glasgow on 9 November 1951. Following a formal handing over ceremony on the dockside the following day, the aircraft were taken to RNAS Abbotsinch to be prepared for service. They were then delivered to RNAS Culdrose, where they equipped No. 778 Squadron commanded by Lt. J. D. Treacher, which had reformed on

5 November 1951. The squadron's task was primarily to evaluate the Royal Navy's planned method of operating aircraft in the AEW role. As the Skyraider had been proved in service with the US Navy, it was decided not to carry out all the usual trials normally demanded for aircraft before they could be accepted for service in the UK, but to restrict trials to catapult take-off and arrested landing, held at RAE Farnborough, and carrier trials on board the newly commissioned HMS *Eagle* during February 1952. The Skyraiders of No. 778 Squadron took part in several operational exercises, including Exercise "Castanets", in conjunction with ships and aircraft when operating from RAF St. Eval.

On 7 July 1952, following the completion of the evaluation task, No. 778 was re-designated as a front line squadron and became No. 849, still at Culdrose and still commanded by the recently promoted Lt. Cdr. J. D. Treacher. When No. 849 reformed there were only four Skyraiders in the UK and it was not until 19 February 1953 that the next batch of 13 aircraft arrived from the USA on board HMS *Perseus,* and this was followed by a further batch on 31 March 1953. However, this lack of aircraft did not prevent the squadron forming 'A' and 'B' Flights during November 1952; although initially the flights were obliged to use aircraft of Headquarters Flight. The plan was to provide a flight of Skyraider AEW Mk 1s for each of the new aircraft carriers then under construction. The arrival of the additional aircraft in 1953 allowed the formation of 'C' Flight in June, 'D' Flight in August and 'E' Flight in December. However, with the introduction of the new technology, angled deck, steam catapult and mirror landing aid, the completion of the carriers was much delayed and resulted in No. 849 Squadron being cut to four operational flights on 5 July 1954. The re-organised squadron consisted of 'A', 'B' and HQ Flights with six aircraft each, and 'C' and 'E' with four aircraft each. By October 1954 it was proving difficult to keep the Skyraiders serviceable because of a shortage of spare engines, and as a consequence 'A' and 'B' Flights were reduced to four aircraft each, and during the same month 'E' Flight became 'D' Flight.

On 19 January 1953, 'B' Flight was detached to RNAS Hal Far, Malta, and operated from there until it

Skyraider AEW Mk 1, WT954/310/J, of 'A' Flight, No. 849 Squadron, taking off from HMS *Eagle* in 1954. (RAF Museum)

returned to Culdrose in November 1953. 'A' Flight was the first to operate at sea, embarking on HMS *Eagle* on 29 January 1953 for a six weeks training cruise and when 'A' Flight next embarked on HMS *Eagle* in June 1953, it was joined by the newly formed 'C' Flight. After two short spells aboard *Eagle*, 'C' Flight was flown out to Hal Far in October 1953, where it replaced 'B' Flight. During its detachment to Hal Far, the flight spent some time embarked in HMS *Glory* in the Mediterranean before returning to the UK in April 1954.

In June 1953, the whole squadron, led by its Commanding Officer, Lt. Cdr. J. D. Treacher, and comprising the HQ, 'A', 'B' and 'C' Flights, took part in the Coronation Review at RNAS Lee-on-Solent.

'A' Flight was the only flight to be totally committed to an aircraft carrier. During its seven years existence with Skyraiders, the flight operated exclusively from HMS *Eagle*, with the exception of an eleven day period of deck landing practice on board HMS *Bulwark* in February 1956, while HMS *Eagle* underwent a short refit at the Devonport Dockyard. The other flights operated from various carriers as required: 'B' Flight from HMS *Eagle*, HMS *Ark Royal* and HMS *Victorious;* 'C' Flight from HMS *Eagle*, HMS *Glory* and HMS *Albion* and 'D' Flight from HMS *Bulwark*, HMS *Albion* and HMS *Centaur*.

When the Suez operation "Musketeer" commenced in November 1956, two flights of Skyraiders were in the Mediterranean: 'A' Flight aboard HMS *Eagle* and 'C' Flight aboard HMS *Albion*. The aircraft of these flights provided a communications service in addition to their basic AEW role for the Allied fleet.

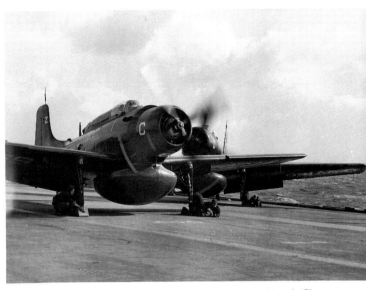

Skyraider AEW Mk 1s, WT969/321/Z and WV180/319/Z, of 'C' Flight, No. 849 Squadron, preparing for take off from HMS *Albion*.

The Skyraider AEW Mk 1 started to be phased out of Fleet Air Arm service in September 1959 with the arrival in 'A' Flight of the first Gannets which were AS Mk 4s for training purposes. The last of this flight's Skyraiders along with the Gannet AS Mk 4s were replaced by Gannet AEW Mk 3s in February 1960. Throughout 1960 the other 849 Squadron Flights converted to Gannets with the exception of 'D' Flight, the last to operate Skyraider AEW Mk 1s at sea, when it embarked in HMS *Albion* on 5 February 1960 for a ten month cruise in the Far East. On *Albion*'s return to Home Waters, 'D' Flight disembarked to Culdrose when it disbanded on 16 December 1960; the aircraft being flown to Abbotsinch on 19 December 1960 for disposal, bringing to an end nearly nine years of first line service.

Twelve of the Skyraiders stored at Abbotsinch were sold to the Swedish Air Board for use as target-tugs towards the end of 1961, and after being refurbished and converted by Scottish Aviation Limited at Prestwick, were delivered to Sweden in Swedish civil markings. With the exception of two Skyraiders retained by the Fleet Air Arm for static display purposes all the other survivors were reduced to spares and scrapped.

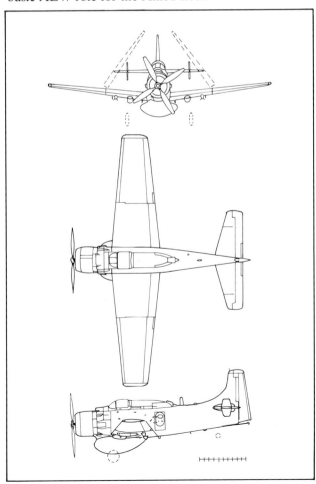

15   Skyraider AEW Mk 1

PRODUCTION

AEW Mk 1 WT097, WT112, WT121, WT761, WT849, WT944 – WT969, WT984 – WT987, WV102 – WV107, WV177 – WV185.

CHAPTER SIXTEEN
Percival Sea Prince

Percival Aircraft Ltd started the design of the P.50 Prince, an eight seat feeder airliner, in 1947. The Prince was to be a fairly basic, conventional high wing monoplane powered by two 500 hp Alvis Leonides radial engines driving three-bladed propellers. The decision to build a prototype was taken in 1947 and shortly after work had started on the prototype, the company laid down a production batch of 10 aircraft. The prototype Prince, bearing the Class B marking G-23-1, made its maiden flight at Luton on 13 May 1948 with Wing Commander H. P. Powell at the controls.

The Royal Navy's interest in the Prince started in 1949 when it was under consideration to meet two Specifications. The first, T.17/19, covered a "flying classroom" to train observers and the second, C.18/49,

First production Sea Prince T Mk 1, WF118/567/BY, of No. 727 Squadron, Brawdy. (R. M. Rayner)

was for a communications aircraft. Percival Aircraft Ltd revised the design of the P.50 Prince to conform to the requirements of the Specification T.14/49. This variant, designated the P.57, met with the approval of the Admiralty who placed a contract on 25 October 1949 for 18 P.57 Sea Prince T Mk 1s and three communications aircraft to Specification C.18/49 and identified as the P.66 Sea Prince C Mk 1.

The Sea Prince C Mk 1 was basically the same as the original civil Prince except for the installation of military equipment and was produced with a standard eight seat layout or with a VIP layout for use as an Admiral's Barge. Several modifications were introduced to meet the requirement for the crew training role. The main change was the introduction of a longer nose to house the radar equipment, but there was also a new longer stroke undercarriage with twin main wheels

Sea Prince T Mk 1, WF127/570/CU, of No. 750 Squadron, Culdrose, 1973. (RNAS Culdrose)

of smaller diameter, rather than the large single wheels of the original Prince. To accommodate the redesigned undercarriage it proved necessary to increase the length of the engine nacelles. The cabin was laid out to carry three students and an instructor along with all the radar and radio consoles necessary for training the students in the techniques of navigation and anti-submarine warfare. A number of the Sea Prince T Mk 1s were equipped with ASH radar with the scanner in a 'thimble'-type radome which protruded from the nose of the aircraft.

The first three Sea Prince C Mk 1s were built using airframes diverted from the first production batch of civil Princes. Consequently the first Sea Prince completed was the C Mk 1, WF137, which was equipped as a special VIP transport and delivered to RNAS Lossiemouth on 22 July 1950 and four days later was transferred to RNAS Lee-on-Solent, where it joined No. 781 Squadron, commanded by Lt. Cdr. D. L. Stirling, the Southern Communications Squadron for use as an Admiral's Barge. The second Sea Prince, WF136, was delivered to the RN Section, Aircraft

Sea Prince C Mk 2, WJ348, used as an 'Admiral's Barge' for the Flag Officer (Air) Home, December 1953.

Handling Squadron at RAF Manby in October 1950 for the compilation of the pilot's notes. The final C Mk 1 was taken on charge by the Joint Services Mission in Washington DC, USA, in July 1951 where it was maintained by the US Navy until its return to the UK in February 1954.

The first Sea Prince T Mk 1, WF118, flew for the first time on 28 June 1951, by which time three further contracts had been placed covering the supply of four Sea Prince C Mk 2s and 23 Sea Prince T Mk 1s. WF118 was used for trials by Percival Aircraft Ltd, RAE Farnborough, and A&AEE Boscombe Down before being delivered to the Fleet Air Arm in April 1954. The third production Sea Prince T Mk 1, WF120, was the first to be delivered direct to the Royal Navy, going to the RN Section, Aircraft Handling Squadron at RAF Manby on 11 December 1951, by way of the Receipt and Despatch Unit (RDU) at RNAS Stretton. The next Sea Prince T Mk 1, WF121, was the first of the type to enter squadron service when in August 1952 it joined No. 744 at RNAS Eglinton, the Trials and Development Squadron of the Naval Air Anti-Submarine School which, under the command of Lt. Cdr. F. E. Cowtan, also operated as the Station Flight.

By the end of November 1953, No. 744 had five Sea Prince T Mk 1s on its strength but shortly afterwards it disbanded at Eglinton. The only front line squadron to

Last production Sea Prince C Mk 2, WM756. (G. A. Jenks)

operate the Sea Prince was No. 831, the Electronic Warfare Squadron commanded by Lt. Cdr. D. K. Blair at RNAS Culdrose, which operated a Sea Prince T Mk 1 in support of the squadron's Gannets and Sea Venoms from July 1962 until the squadron disbanded in 1966. In addition, each of the RNVR anti-submarine squadrons operated a Sea Prince T Mk 1 in support of its operational aircraft which were Fireflies, Avengers or Gannets, until they were disbanded in early 1957. The Sea Princes were used by the squadrons for observer training and communications purposes.

The four Sea Prince C Mk 2s were basically T Mk 1 airframes fitted out for the communications role. These were completed during 1953 and each was to spend some time as an Admiral's Barge and usually operated

under the auspices of No. 781 Squadron. The first Sea Prince to be taken out of service was a C Mk 2, WJ348, which was broken up for spares at RNAS Donibristle in January 1958, and this was followed by another C Mk 1, WF136, which was struck off charge in November 1965. The remaining two C Mk 1s and the second C Mk 2 were struck off charge in 1969, leaving the final two C Mk 2s to carry on for a few more years, but both these had disappeared from the scene by November 1973.

The prime user of the Sea Prince T Mk 1 was No. 750 Squadron which had reformed in April 1952 at St. Merryn, equipped with Barracudas and Ansons, as part of the Royal Navy's Observer School. In February 1953 the Barracudas and Ansons were replaced by Fireflies and Sea Princes and shortly afterwards in November 1953 the Squadron moved to RNAS Culdrose where it became the Observer and Air Signal School. In March 1955 the Fireflies were transferred to No. 796 Squadron, leaving No. 750 to concentrate on the basic training of observers with the Sea Princes. The squadron was operating a full complement of nine Sea Prince T Mk 1s when in October 1959 it was transferred to Hal Far in Malta and continued to operate there until July 1965 when the Sea Princes were flown to Lossiemouth. On 26 September 1972 the squadron was transferred to Culdrose and as the Sea Princes were coming to the end of their fatigue lives at this time a refurbishment programme was started by the British Aircraft Corporation. This included a series of fatigue remedial modifications to extend the life of the surviving Sea Princes until a suitable replacement could be found. Subsequently the BAe Jetstream was selected and the Jetstream T Mk 2 started to replace No. 750 Squadron's Sea Princes in October 1978, with the last Sea Prince being phased out of service in May 1979 some 29 years after the delivery of the first to the Royal Navy.

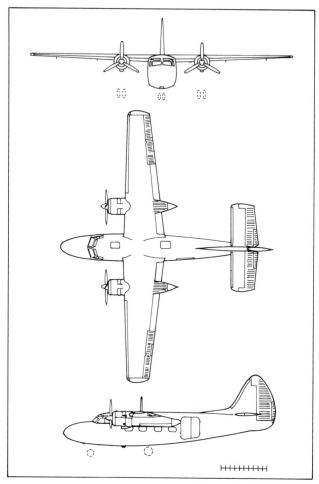

16   Sea Prince T Mk 1

PRODUCTION

T Mk 1   WF118 – WF133, WF934, WF949, WM735 – WM742, WP307 – WP321.
C Mk 1   WF136 – WF138.
C Mk 2   WJ348 – WJ350, WM756.

CHAPTER SEVENTEEN

Gloster Meteor

Following the success of Britain's first turbo-jet powered aircraft, the Gloster E.28/39, the Air Ministry decided that the next stage of development using this revolutionary new method of propulsion would be a fighter. Due to the limited amount of thrust produced by the early turbo-jet engines the Gloster design team, led by their chief designer George Carter, produced, during the mid 1940s, the design of a twin-engined fighter with an armament of six 20mm cannons. In November 1940 the Air Ministry issued Specification F.9/40 which had been written around the Gloster design. Twelve prototypes, DG202/G to DG213/G, were ordered, but the quantity was subsequently adjusted and only eight were actually built.

Meteor F Mk III, EE337, being manhandled on board HMS *Implacable* during carrier trials in June 1948. (RAF Museum)

As work progressed on the prototypes, the type was officially named Meteor in February 1942. The first Meteor to fly was the fourth prototype DG206/G powered by two Halford H.1 engines which was flown at Barford St. John by Michael Daunt on 5 March 1943.

The fleet of prototypes was fitted with a variety of engines including the Whittle W.2b engine which when subsequently built by Rolls Royce was named the Welland, and was later developed to become the Derwent. The de Havilland/Halford H.1 was installed in two of the prototypes, and this engine was developed into the Goblin which was to power the de Havilland Vampire. The third engine used was the Metropolitan Vickers F.2 engine, which was later developed by Armstrong Siddeley Motors into the Sapphire. This had an axial compressor, unlike the other two, which were of the centrifugal compressor type.

Meteor T Mk 7, VW447/935/LM, of RNAS Lossiemouth's Station Flight. (B. J. Lowe)

The Meteor was ordered into production, and EE210/G, the first Meteor F Mk 1, powered by the Rolls Royce W.2b/23c Welland flew for the first time on 12 January 1944. By the time 20 Meteor F Mk1s had been built the new Derwent engine was available and the next production version, the F Mk 3, powered by this engine had been put into quantity production. The F Mk 2 designation had been allotted to the Halford H.1 powered Meteor, but as these engines had been selected for the Vampire it was decided not to put the Meteor F Mk 2 into production and only one prototype, DG207/G, was brought up to F Mk 2 standard. In total, 210 production Meteor F Mk 3s were delivered between December 1944 and April 1946.

The Royal Navy's first involvement with the Meteor started on 11 August 1945 when the first prototype, DG202/G, was flown to RNAS Abbotsinch by Gloster test pilot Eric Greenwood. From there it was transported by lighter to HMS *Pretoria Castle* for deck handling trials only, no take-offs or landings being attempted. Although there was no real prospect of the Meteor being developed into a ship-borne fighter, the Admiralty was convinced that carrier trials with a modified Meteor would provide useful data on operating a twin jet aeroplane from a carrier.

As part of this programme Lt. E. M. Brown carried out an assessment of the suitability of the Meteor for operation on a carrier deck, using an F Mk 3, EE476. Following the satisfactory completion of this assessment, the Gloster Aircraft Company modified two Meteor F Mk 3s, EE337 and EE387, at their Moreton Valence factory. All unnecessary equipment, including the armament, was removed from the aircraft and Sea Hornet 'A' frame type arrester hooks were fitted under the rear fuselage. To absorb the shock of arrested landings, the fuselage was strengthened and a new stronger undercarriage was also introduced. The main undercarriage doors were removed because of the danger of them tangling with the arrester wires on landing, and because of this the maximum flying speed was limited to 250 knots (288 mph). In addition improvements were made to the braking system and the more powerful Derwent V engines were installed.

After completing a programme of airfield dummy deck landings at the A&AEE, Boscombe Down, EE387 carried out initial carrier trials on HMS *Illustrious*. At about the same time EE337 was delivered to RAE Farnborough for assessment and on 8 June 1948 Lt. E. M. Brown landed EE337 aboard HMS *Implacable* which was steaming in the English Channel. During these trials, a total of 32 landings were carried out. Although Lt. Brown considered that an increase in flap area would have improved the Meteor's deck performance he found the take-off and landing relatively straightforward. Following the trials, EE337 served with No. 703 Squadron, the Royal Navy's service trials unit at Ford from August 1948 until 1952. It is understood that both of the Royal Navy's Meteor F Mk 3s were scrapped during 1955.

In 1947, with the increasing demand for Meteor fighters for the RAF and various overseas air forces, the Gloster Aircraft Company realised that there was a requirement for a two-seat trainer version. Work was immediately put in hand to design a Meteor trainer as a private venture and the design proved to be a relatively simple conversion of the contemporary production version, the Meteor F Mk 4. The first Meteor trainer designated the T Mk 7 was a conversion of the company's F Mk 4 demonstrator, G-AIDC which had been damaged in a landing in Belgium during a European sales tour. Towards the end of 1947, a new fuselage, which accommodated the instructor and student seated in tandem under a long canopy, was fitted to G-AIDC; damage to other parts of the airframe was repaired and the aircraft emerged with a new civil registration, G-AKPK and flew for the first time on 19 March 1948. The Meteor T Mk 7 proved to be a considerable success and was ordered in large quantities for the RAF.

With plans to introduce the Supermarine Attacker, followed by the Hawker Sea Hawk into operational service, the Royal Navy decided to update its training organisation by re-equipping with the next generation of trainers. As part of this modernisation programme the Royal Navy ordered a small batch of Meteor T Mk

Meteor U Mk 15, VT268/656, of No. 728B Squadron, Hal Far.

7s which were diverted from the RAF production batches and between 1948 and 1953 a total of 35 aircraft were delivered to the Fleet Air Arm. The Meteor T Mk 7s were used by a large number of second line units mainly for training and sometimes for communications duties. The largest single user of the type was No. 759 Squadron which ran a jet conversion course at RNAS Culdrose and operated them along with Sea Vampires from September 1952 until April 1954. At its peak in 1953, No. 759 was operating 14 Meteor T Mk 7s.

Several other second line squadrons operated T Mk 7s, in some cases only a single aircraft for communications or instrument rating purposes. It was No. 728 Squadron based at Hal Far in Malta that was to use the Meteor T Mk 7 for the longest period, receiving its first aircraft in February 1955 and continuing to operate the type until it disbanded in May 1967. This squadron operated a variety of aircraft on a wide range of duties including communications, target-towing, and search and rescue and one would assume that, in addition to their use as trainers in support of the target-towing Meteor TT Mk 20s, the Meteor T Mk 7s were used as squadron hacks. Like many other Fleet Air Arm aircraft, the Meteor T Mk 7 ended its service career in

Meteor U Mk 16, WF716/658, of No. 728B Squadron, Hal Far.
(A. E. Hughes)

March 1971 with the civilian operated (Airwork Ltd) Fleet Requirements Unit at Hurn.

The feasibility of using Meteors as target drones was considered in the early 1950s and in 1954 a Meteor T Mk 7, VW413, was modified to act as a trials aircraft and flew for the first time under radio control, although still carrying a pilot, on 17 January 1955. Completion of the trials coincided with the RAF replacing its Meteor F Mk 4s with the later F Mk 8s, making a large number of aircraft available for conversion into pilotless target drones. Initially, 92 Meteor F Mk 4s were converted into Meteor U Mk 15s and later 108 Meteor F Mk 8s were converted into Meteor U Mk 16s. Most of these aircraft were operated by the RAE at Llanbedr for use in the development and trials of ground-to-air missiles. However, by the end of 1958 it was apparent that the Firefly U Mk 8 and U Mk 9 aircraft operated by No. 728B Squadron at Hal Far, Malta, would need to be supplemented for the AWA Seaslug trials being carried out from HMS *Girdle Ness* in the Mediterranean, and to meet the demand 17 Meteor U Mk 15s were transferred to the Royal Navy. These aircraft were delivered to No. 728B in July 1959, and were followed by four Meteor U Mk 16s in October 1960. In December 1961, with the Seaslug trials completed, No. 728B disbanded and the surviving Meteors were transferred to the RAE at Llanbedr.

The final and perhaps most extensively used of the Royal Navy's Meteors was the TT Mk 20 which was

Meteor TT Mk 20, WM159/040, of the Airwork FRU, Hurn, 1967.

developed from the NF Mk 11 to meet a requirement for a modern high speed target tug to replace the Short Sturgeon TT Mk 3. In 1956, a design study was started for conversion of the NF Mk 11 to the target-towing role and WD767 was allotted to Armstrong Whitworth Aircraft Company for the development programme. The conversion consisted of the introduction of an ML Aviation Type G wind-driven winch, mounted above the starboard wing between the fuselage and the engine

nacelle, providing 6100ft of cable. As many as four sleeve targets could be carried in a fairing under the fuselage centre section, and a tail guard wire was fitted to prevent the control surfaces from being fouled by the towing cable. The rear crew member was responsible for the target winch and for operating the cable cutter which jettisoned the target. The airborne interception radar and armament were deleted and the VHF radio was replaced by a UHF set. After modification, WD767 made its first flight as a Meteor TT Mk 20 on 5 December 1956 and, following trials at Bitteswell and the A&AEE, Boscombe Down, a contract was placed with AWA for the conversion of 18 Meteor NF Mk 11s to TT Mk 20 standard.

Additional aircraft were converted by the Gloster Aircraft Company and the Royal Navy Air Yard at Sydenham, but a number of these were delivered to the RAF, and in fact only 29 aircraft saw service with the Fleet Air Arm. The Meteor TT Mk 20 was first used by No. 728 Squadron at Hal Far, Malta, when it started to replace its Sturgeon TT Mk 3s in March 1958, and remained with the squadron until disbanded in May 1967. A small number of Meteor TT Mk 20s were used on fleet requirements duties by No. 700 Squadron at Yeovilton from December 1959 until it disbanded in July 1961. From that time on the squadron's fleet requirements work was taken over by the Fleet Requirements Unit operated by Airwork at Hurn which had been operating Meteor TT Mk 20s since May 1958, and continued to do so until March 1971 when the last were replaced by Canberra TT Mk 18s.

PRODUCTION

F Mk 3   EE337, EE387.
T Mk 7   VW446, VW447, VZ645-VZ648, WA600, WA649-WA651, WL332-WL337, WL350-WL353, WS103-WS117.
U Mk 15   RA375, RA387, RA479, VT104, VT107, VT110, VT196, VT243, VT258, VT268, VT282, VT291, VT310, VW258, VW276, VZ415, VZ417.
U Mk 16   WE932, WF716, WK870, WL127.
TT Mk 20   WD585, WD592, WD606, WD610, WD612, WD641, WD643, WD645, WD646, WD649, WD652, WD657, WD678, WD706, WD711, WD780, WD785, WM147, WM148, WM151, WM158-WM160, WM181, WM230, WM242, WM255, WM260, WM292.

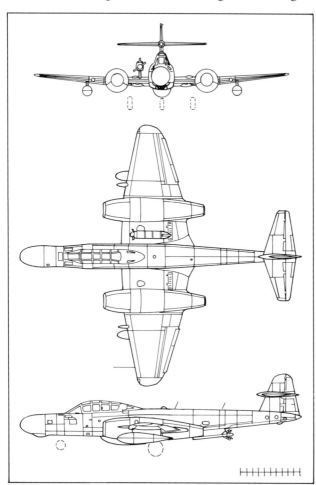

17   Meteor TT Mk 20

CHAPTER EIGHTEEN
Westland/Sikorsky Whirlwind

Although the Dragonfly had demonstrated the value of the helicopter to the Fleet Air Arm, it had also shown that its capacity and performance severely limited its usefulness as an operational aircraft, and consequently it was used only for training, air-sea rescue and communications purposes. With this in mind Sikorsky had designed the S 55, a helicopter which was significantly larger with a more powerful engine and consequently capable of carrying a greater payload.

The prototype S 55 made its first flight in the USA on 10 November 1949 and Westland, realising the potential of this new development, started negotiations with Sikorsky the following year for a licence to build the S 55 in the UK and this was obtained in November 1950. Following the procedure adopted with the Dragonfly,

Westland initially purchased an S 55, which by now had been named the Whirlwind in the USA. This was delivered early in 1951 and was registered by the company as G-AMHK. Following trials by Westlands and an evaluation by the RAF in 1952 the aircraft was sold abroad in September 1953.

Before Westland could get production of the Whirlwind under way, a batch of 25 were ordered from Sikorsky for the Royal Navy and these were supplied under the Mutual Defence Aid Programme (MDAP). The first ten of these Sikorsky built Whirlwinds were HRS-2s used for rescue and general duties and were re-designated HAR Mk 21s for service with the Fleet Air Arm. The remaining 15 were HO4S-3s equipped with active dunking sonar for anti-submarine duties, and these were re-designated HAS Mks 22s.

All ten of the Whirlwind HAR Mk 21s were delivered during 1952 and were used to equip No. 848

Whirlwind HAR Mk 21, WV198/K, of No. 848 Squadron, during anti-terrorist operations in Malaya. (via E. A. Myall)

Squadron, commanded by Lt. Cdr. S. H. Suthers DSC DFC, at RNAS Gosport. This squadron reformed on 29 October 1952 and was the first Fleet Air Arm front line helicopter squadron. The squadron was transported to Singapore in December 1952 aboard HMS *Perseus,* and was used in support of British Forces during the Malayan Emergency. After working-up at RNAS Sembawang, three Whirlwinds were based at Kuala Lumpur with the remainder staying at Sembawang. They soon demonstrated their value by being able to deploy large numbers of troops into otherwise inaccessible jungle areas in relatively short times, and were also very useful for casualty evacuation duties. For its operations in Malaya No. 848 Squadron was awarded the Boyd Trophy for 1953. The squadron was to continue operating on the Malayan Peninsula until it was disbanded at Sembawang in December 1956.

With the arrival of the anti-submarine Whirlwind HAS Mk 22s from the USA, No. 706 Squadron, under the command of Lt. Cdr. H. Phillips, was reformed on 7 September 1953 at RNAS Gosport as an Anti-Submarine Helicopter Trials Unit equipped with eight Whirlwind HAS Mk 22s and two Hiller HT Mk 1s. During the period up to 15 March 1954 the squadron carried out trials with the sonar, and for three weeks early in 1954 operated aboard HMS *Perseus.* When No. 706 disbanded on 15 March 1954 it was immediately re-designated No. 845, and became the first front line helicopter anti-submarine squadron in the Fleet Air Arm. On 21 April 1954 the squadron embarked in HMS *Perseus,* sailed for RNAS Hal Far, and was to remain operating in the Mediterranean for the next 18 months; either from Hal Far or from one of the British carriers operating in the area.

The only other front line squadron to use the Whirlwind HAS Mk 22 was No. 848, which had previously used the Whirlwind HAR Mk 21s. This squadron reformed at RNAS Hal Far on 14 October 1958 with five Whirlwind HAS Mk 22s as the Amphibious Warfare Trials Unit and, serving with 45 Commando, became the first Royal Marine Commando helicopter squadron. The Whirlwind HAS Mk 22s were phased out of front line service in November 1959, being replaced by the Whirlwind HAS Mk 7. A small number of the American built Whirlwinds continued to operate throughout the 1960s with several of the second line squadrons until they finally disappeared from the scene, the last one operated by No. 781 being replaced

Whirlwind HAS Mk 22, WV224/958, of the SAR Flight, Hal Far.
(A. E. Hughes)

by a Westland Wessex in March 1970.

Westland's first WS 55 Whirlwind was allocated the civil registration G-AMJT but by the time it flew for the first time on 15 August 1953 it had been allotted to the Royal Navy and given serial number XA862. Initially it had been planned to name the WS 55 the Cranefly, but common sense prevailed and the name used by the Americans was adopted. The first batch of Westland built Whirlwinds were purely anglicised Whirlwind HRS-2s powered by the 600hp Pratt and Whitney Wasp R-1340-40 engine. Ten of the Series 1 Whirlwinds were built for the Royal Navy as the HAR Mk 1 along with ten for the RAF as the HAR Mk 2. Deliveries of the HAR Mk 1s started in July 1954 and in September five were transported to Sembawang to support No. 848 Squadron's Whirlwind HAR Mk 21s in the search and rescue role. The HAR Mk 1s were only on the strength of No. 848 for a short time and were withdrawn in June 1955. During the conversion of HMS *Protector* into an ice patrol ship a helicopter landing pad was introduced and in July 1955 a ship's flight was formed with two Whirlwind HAR Mk 1s and was to continue operating these on regular patrols in the South Atlantic until May 1966 when they were replaced by Whirlwind HAR Mk 9s. In December 1964 No. 829 Squadron took over the responsibility for HMS *Protector*'s ship's flight, which made it the last squadron to operate the Whirlwind HAR Mk 1.

The next version of the Whirlwind for the Royal Navy was the HAR Mk 3 which was powered by the 700hp Wright R-1300-3 Cyclone engine. An order was placed for 29 HAR Mk 3s, intended primarily for the search and rescue role and which started to replace the Dragonflies of the ship's flights in September 1955 with deliveries to HMS *Ark Royal* and later to *Albion, Warrior, Bulwark* and *Eagle.* All of these aircraft were phased out of service in October 1957 with the exception of those on HMS *Eagle* which were retained until April 1959. It was however No. 705 Squadron that was to use Whirlwind HAR Mk 3s longer than any other unit, receiving the first aircraft in November 1955 and operating the type, along with a variety of other helicopters until they were finally phased out of service in February 1966.

Although used by seven second line squadrons the HAR Mk 3 had only limited use by front line squadrons. The first of these, No. 845, commanded by Lt. Cdr. J. C. Jacob, operated two in support of its Whirlwind HAS Mk 22s from the time it was reformed on 14 November 1955 until re-equipped with the later Whirlwind HAS Mk 7s in August 1957. No. 815 was the only other front line squadron to operate HAR Mk 3s, re-equipping with them in January 1959, when the squadron's HAS Mk 7s were grounded by engine faults. It operated them until August 1959 when the squadron was relegated to second line status and became No. 737 Squadron at RNAS Portland.

To meet the Royal Navy's requirement for an anti-submarine helicopter capable of carrying all the necessary detection equipment and also the weaponry to destroy enemy submarines, Westland decided that a more powerful engined Whirlwind would fit the bill. The engine selected was the new 750hp Alvis Leonides Major Mk 5 and the first installations of this engine were made in 1955 on two Whirlwind HAR Mk 3s, XG589 and XG597, which were converted on the production line into the prototype Mk 5s, the first of

Flight of four Firefly T Mk 7s, including WJ192/302/MF, WJ168/300/MF and WM765/303/MF, of No. 750 Squadron, St. Merryn.
(RAF Museum)

tained for a few months until the squadrons reached operational status, when they were replaced by Firefly Mk 5s or 6s.

Fairey realised at the end of 1948 that there was almost certainly going to be a gap on the production line between the end of the Firefly and the beginning of Gannet production. Looking for a means to cover this anticipated lean period, the company came to the conclusion that there could be a requirement for an aircraft to bridge the operational gap between the Firefly AS Mk 6 and the Gannet, which was then early in its development programme. Their answer to this possible requirement was a development of the Firefly AS Mk 6 which introduced an additional observer and used the new British sonobuoy equipment which was under development at that time. Fairey believed that this new Firefly would be relatively simple to develop, allowing the aircraft to be delivered to service quickly and at low cost. The Admiralty agreed that this new Firefly development would bridge the gap between the Firefly AS Mk 6 and the Gannet AS Mk 1 and, perhaps more importantly to them, it would provide cover in the event of there being any delay in the Gannet entering service. It was also believed that the new Firefly would be more suitable than the larger and heavier Gannet for operation on the Royal Navy's smaller carriers and consequently could remain in service for some time after the Gannet had entered service. In the event this new Firefly, identified as the AS Mk 7, proved to be a rather more significant re-design of the Firefly Mk 6 than was originally envisaged. The provision for the extra crew member and the related additional equipment resulted in a much heavier aircraft than expected, requiring an increase in wing area, together with a more powerful engine to meet the performance requirements.

In April 1950 the company received Specification M.101P which set the production standard for the Firefly AS Mk 7. This was followed by a production order for 54 aircraft, soon to be increased to 80.

Deliveries were to commence in April 1951 and be completed within twelve months. To meet the specification, the 1965hp Rolls Royce Griffon 59 was selected to power the Firefly Mk 7, which required the re-introduction of a deep 'beard' radiator rather larger than that used on the earlier marks of Firefly. The rear observer's cockpit was enlarged to take the extra crew member and his equipment and covered with a large bulged canopy similar to the one designed for the Gannet. A wing of increased span and changed planform was introduced, along with a large angular fin and rudder to counteract the destabilising effect of the new radiator system.

The first pre-production aircraft, WJ215, flew on 22 May 1951, by which time an additional 206 Firefly AS Mk 7s had been ordered. Faireys were having trouble in reducing the weight to an acceptable level, and trials at the A&AEE at Boscombe Down resulted in a highly critical report, particularly in regard to handling qualities at low speeds. Despite all the problems being encountered, a further 51 aircraft were ordered, bringing the total to 337. By mid 1952 it had become evident that the Firefly AS Mk 7 would never be a satisfactory carrier-borne operational aircraft, and consequently it was relegated to a training role, becoming the Firefly T Mk 7. The production orders were cut and eventually only 151 T Mk 7s were built. The Firefly AS Mk 7's intended role as a stopgap anti-submarine aircraft was met partly by extending the operational life of the Firefly AS Mk 6s and by using the Grumman Avenger AS Mk 4s and 5s that had been delivered to the Fleet Air Arm from the USA under the Mutual Defence Aid Programme (MDAP).

The Firefly T Mk 7 became the basic equipment of three of the Fleet Air Arm's training units, namely No. 719 Squadron, the Naval Air Anti-Submarine School at Eglinton from March 1953 to June 1956, No. 750 Squadron, the Observer and Air Signal School at St. Merryn and Culdrose from April 1953 to March 1955, and No. 796 Squadron, the Observer's School at Eglinton from June 1953 to December 1957. In addition, the T Mk 7s were used by No. 765 Squadron, the Piston Engine Pilot's Pool at Culdrose from June

Two Firefly U Mk 9s, including VT485/592, of No. 728B Squadron, Hal Far, Malta.

1953 to December 1957, and No. 1840 RNVR Squadron to train observers for the Gannet AS Mk 1 aircraft.

For the Coronation Review of the Royal Navy on 15 June 1953, some 64 Fireflies of all marks from ten squadrons took part in the flypast, making it the largest contingent of any type that took part in the event.

The Firefly T Mk 7 was, however, not quite the end of the Firefly saga, for in 1952 there was a requirement for pilotless target drones to be used for missile development programmes. Initially six Firefly T Mk 7s were allocated to the programme and were modified by the introduction of radio control gear, which allowed the aircraft to be flown from the ground; although a

facility was retained for it to be flown normally by a pilot. Pods were fitted to the wingtips containing cameras that recorded the trials. These aircraft, identified as the Firefly U Mk 8, were finished in a bright red and cream paint scheme to make them easy to see from the ground. The first Firefly U Mk 8, WM810, flew on 30 December 1953. In addition to the initial batch of six Mk 7s that were converted to U Mk 8s, there were 34 new build aircraft. An increased demand for target drones resulted in the conversion of 40 Firefly Mk 5s using the same equipment as had been used for the U Mk 8, under the designation U Mk 9. The Firefly target drones were mainly used by the RAE at Llanbedr, and No. 728B Squadron at Hal Far in Malta.

The final total of Firefly FR Mk IV to U Mk 8 production amounted to 776 airframes, which included the aircraft that were supplied to Canada and Australia. By the time No. 728B Squadron lost their last U Mk 9 in November 1961, Fireflies had recorded some 18 years' continuous service in the Fleet Air Arm.

Firefly U Mk 9, WB257/591, of No. 728B Squadron, Hal Far.
(A. E. Hughes)

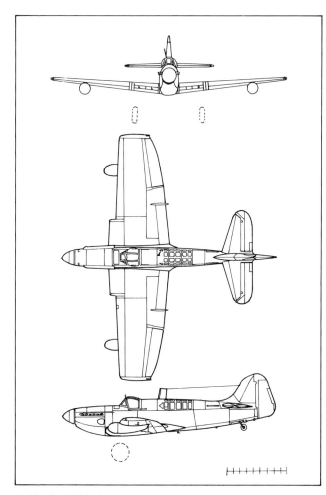

8 Firefly AS Mk 6

Production

Mk IV TW687 – TW699, TW715 – TW754 (originally ordered as Mk 1s), VG957 – VG999, VH121 – VH144.

Mk 5 VT362 - VT381, VT392 – VT441, VT458 – VT504, VX371 – VX396, VX413 – VX438, WB243 – WB272, WB281 – WB316, WB330 – WB382, WB391 – WB440.

Mk 6 WB505 – WB510, WB516 – WB523, WD824 – WD872, WD878 – WD926, WH627 – WH632, WJ104 – WJ121.

Mk 7/8 WJ146 – WJ174, WJ187 – WJ209, WJ215 – WJ216, WK348 – WK373, WM761 – WM779, WM796 – WM832, WM855 – WM899, WP351 – WP354.

CHAPTER NINE
De Havilland Sea Vampire

The de Havilland DH100 was designed to meet Air Ministry Specification E.6/41 for an experimental aircraft powered by a single 2700lb thrust de Havilland Goblin centrifugal turbo-jet engine, and with the potential to be developed into a single-seat fighter. The de Havilland Aircraft Company received an Instruction to Proceed in May 1942 for the manufacture of three prototypes. The aircraft had a twin boom layout with the wings, booms and tail surfaces of conventional light alloy construction. The fuselage, however, utilised Mosquito technology, being made in two halves from a plywood/balsa sandwich. Each half was equipped with all the basic fittings before being joined together along the centre line. Initially the DH100 was allocated the type name "Spider Crab" but fortunately this was subsequently replaced by the name "Vampire", a much more appropriate name for a fighter.

The first prototype Vampire, LZ548/G, made its maiden flight at Hatfield Aerodrome on 20 September

Vampire LZ551 taking off from HMS *Ocean* during the first carrier trials by a jet aircraft on 3 December 1945. (B.Ae.)

Sea Vampire F Mk 20 VV138 going down on the lift to the hangar deck of HMS *Theseus* during June 1950. (B. J. Lowe).

1943 in the hands of the chief test pilot Geoffrey de Havilland Jnr and was soon joined by the other two prototypes, LZ551/G and MP838/G. A production order for 120 Vampires was placed in mid 1944 but, as the production facilities at Hatfield were fully committed to the Mosquito, production of the Vampire was started at the English Electric factories at Preston and Salmesbury.

The Fleet Air Arm started to take an interest in the possibilities of jet aircraft in May 1944, initially with the Gloster Meteor and later the Vampire. However, it was not until May 1945 that Lt. E. M. Brown started a deck landing assessment of the Vampire at Hatfield using the second prototype, LZ551/G, which, to meet the requirement for carrier operation, had had its flaps and

airbrakes increased in area. The assessment showed that the aircraft would unstick at 87 knots (100 mph) in a distance which would permit a free take off from a carrier deck and with a landing speed of 91 knots (105 mph) it was considered that the Vampire would be suitable for carrier operations.

When LZ551/G returned to the RAE Farnborough in October 1945 it had been fitted with a V-frame arrester hook attached to the rear of the wing roots and which retracted into a pen-nib type fairing protruding aft from above the jet pipe. Problems with the hook during trials with the arrester gear on the runway at Farnborough resulted in the aircraft being delivered to the de Havilland factory at Christchurch for modifications. While it was there the opportunity was taken to introduce the new 3100lb thrust Goblin 2 engine, and also a bubble canopy hood to improve the pilot's visibility. On its return to Farnborough in November the arrested landing trials were successfully completed, after which the aircraft was flown to RNAS Ford by Lt. Brown to carry out Aerodrome Dummy Deck Landings (ADDLs).

On 3 December 1945 Lt. Brown flew LZ551/G out to HMS *Ocean* sailing off the Isle of Wight. Despite the fact that the ship was pitching and rolling rather excessively for initial carrier trials, the Vampire accomplished the first ever deck landing by a jet powered aircraft. During the fourth landing the bottom edge of the flaps struck an arrester wire, damaging the flap hinge brackets. While this damage was being repaired, the flaps were also reduced in area to increase ground clearance, enabling the trials on HMS *Ocean* to be completed satisfactorily.

The trials confirmed that these early turbo-jet engines were really too slow in responding to the engine controls and the acceleration did not compare at all

A flight of four Sea Vampire T Mk 22s, XA169/611/LM, XG746/601/LM, XA168/603/LM and XA163/600/LM of No. 736 Squadron, Lossiemouth in 1957. (B.Ae.)

Whirlwind HAR Mk 1, XA865, landing at the South Bank Heliport in London. (RAF Museum)

the Series 2 Whirlwinds. These two aircraft were used entirely for development purposes and the Mk 5 was not ordered into production. The next production version was the HAS Mk 7 and orders were placed for 129 of these anti-submarine warfare (ASW) helicopters to replace the Fairey Gannet. Using the same engine as the Mk 5, the HAS Mk 7 was equipped with radar and a dunking sonar. Armament comprised either a Mk 30 or Mk 44 lightweight homing torpedo, or bombs or depth charges carried in a ventral weapons bay although unfortunately the Whirlwind was still unable to carry both the sonar and the weapons.

Deliveries of the HAS Mk 7s to the Fleet Air Arm started early in 1957 with a number of aircraft initially going to the Whirlwind intensive flying trials unit, No. 700H Flight commanded by Lt. Cdr. J. G. C. Williams, at RNAS Lee-on-Solent. The flight formed on 18 March 1957 and after successful completion of the trials, disbanded on 26 September 1957.

During June 1957 No. 845 under the command of Lt. Cdr. H. M. A. Hayes became the first front line squadron to use HAS Mk 7s when it re-equipped with eight of them at Lee-on-Solent. In August of the same year it embarked on HMS Bulwark for a series of exercises. With the decision having been taken to replace the Fleet Air Arm's fixed wing anti-submarine aircraft with helicopters, front line squadrons continued to re-equip with HAS Mk 7s and during 1958 three squadrons re-equipped with the HAS Mk 7 at RNAS Eglinton: No. 820 in January, No. 824 in April and No. 815 in September. Unfortunately, from the beginning of its service career the Whirlwind HAS Mk 7 suffered serious engine problems as a result of which a number of aircraft were lost and during mid 1959 the Fleet Air Arm's entire HAS Mk 7 fleet was grounded until the problems were remedied.

In all a total of seven front line anti-submarine squadrons operated HAS Mk 7s and when their career as anti-submarine aircraft was over, a number had the equipment stripped out and, during the 1960s were used as transports for up to eight fully equipped Royal Marine Commandos, operating from the Commando Carriers. Following the success of the Whirlwinds in the troop transport role with No. 848 Squadron in Malaya, they had further demonstrated their worth during

Operation "Musketeer", the Anglo/French landings at Suez in November 1956, when No. 845 equipped with HAS Mk 22s converted to the trooping role together with the Whirlwinds and Bristol Sycamores of the RAF/Army Joint Helicopter Development Unit had been used to take No. 45 Commando, Royal Marines ashore at the beginning of the land operations.

During the early 1960s three commando squadrons were formed, equipped with the modified HAS Mk 7s. No. 848 commanded by Lt. Cdr. B. M. Tobey, was the first, re-equipping with 16 HAS Mk 7s at Worthy Down in November 1959 and spending some time in the Far East operating from HMS Bulwark, before disbanding at Culdrose in July 1963. The second was No. 846, under the command of Lt. Cdr. D. F. Burke MBE, which had reformed at Culdrose in May 1962 with six HAS Mk 7s, embarked in HMS Albion in September and sailed for the Far East, there to be involved in providing support for operations against terrorists in Brunei; a role that was to continue until October 1964 when the squadron sailed in HMS Bulwark to Singapore where it was disbanded. For its work in Brunei the squadron was awarded the Boyd Trophy for 1963. The last of the Whirlwind commando squadrons was No. 847 commanded by Lt. Cdr. G. A. Andrews which had reformed in May 1963 at Culdrose from part of No. 848 with twelve HAS Mk 7s. The squadron remained at Culdrose where it was used for training pilots for operations in the Far East, although for a period 'B' Flight embarked in HMS Bulwark for exercises in the Far East. No. 847 disbanded in December 1964 at Culdrose, and was the last of the Fleet Air Arm's front line squadrons to be equipped with the Whirlwind, although subsequently a few front line squadrons did have one or two in addition to their main equipment, generally for communications duties.

Whirlwind HAS Mk 7s were used by a number of Ship and Station Flights, with the ship's flight on HMS Albion being the first to use the type, receiving its first aircraft in June 1958. HMS Ark Royal's ship's flight proved to be the last to operate Whirlwinds in September 1966. Four station flights operated HAS Mk 7s with RNAS Culdrose being the first to use them in December 1959 and RNAS Lossiemouth the last in September 1972. This was however not quite the end of the HAS Mk 7s as a small number continued in second line service with No. 705 Squadron until December

Whirlwind HAS Mk 3, XG577/301, of No. 815 Squadron operating its dunking sonar. No. 815 Squadron was the only first line squadron to operate the HAS Mk 3. (via E. A. Myall)

Whirlwind HAS Mk 7, XL844/295/R, of No. 820 Squadron, HMS *Ark Royal,* at Hal Far, 1960. (A. E. Hughes)

1974. As mentioned earlier, the Mk 5 never went into production, but in 1958 the second prototype, XG597, was modified to take the General Electric T-58 turbine engine and flew for the first time on 28 February 1959. The engine installation proved to be relatively simple, the only external differences being a lengthened nose and a large exhaust duct.

Subsequently the de Havilland Engine Company started to build the T-58 engine under licence as the Gnome H.1000 and it was with this engine that the Royal Navy decided somewhat belatedly to re-engine 17 of their Mk 7s, which converted them into HAR Mk 9s. The first HAR Mk 9 flew for the first time on 15 October 1965 and the type started to enter service at the beginning of 1966 with the first joining the ship's flight of HMS *Hermes* in January 1966 for search and rescue duties. HMS *Protector*'s ship's flight was the next to receive Mk 9s and when *Endurance* replaced *Protector* in 1968, the ship's flight was transferred, and the Whirlwinds were retained by *Endurance* until they were replaced by Westland Wasps in September 1976. The other operators of the Mk 9s were the station flights at Culdrose, Brawdy, Lossiemouth and Lee-on-Solent and it was the Lee-on-Solent station flight that proved to be the final operator of the Whirlwind, taking its last HAR Mk 9 out of service in March 1977 some 25 years after the first Whirlwind had entered service with the Fleet Air Arm.

Four Whirlwind HAR Mk 9s, including XL899/587/CU and XL839/588/CU, of the SAR Flight, Culdrose, flying over St Michael's Mount. (RNAS Culdrose)

PRODUCTION

HAR Mk 21 WV189 – WV198.
HAS Mk 22 WV199 – WV205, WV218 – WV225.
HAR Mk 1 XA862 – XA871.
HAR Mk 3 XG572 – XG588, XJ393 – XJ402.
HAS Mk 7/9 XG589 – XG597, XK906 – XK912, XK933 – XK945, XL833 – XL854, XL867 – XL884, XL896 – XL900, XM660 – XM669, XM683 – XM687, XN258 – XN264, XN297 – XN314, XN357 – XN362, XN379 – XN387.

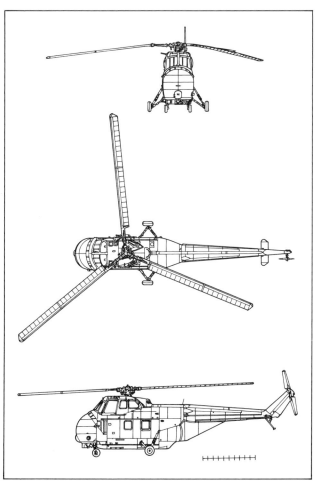

18   Whirlwind HAS Mk 7

CHAPTER NINETEEN

Hawker Sea Hawk

Early in 1944, with more reliable turbojet engines under development, Hawker's turned their attention to the problems associated with the production of jet powered aircraft. Their early design work concentrated on a single jet engined fighter bomber replacement for the F2/43 Fury. However it was not until September 1945 that Hawker's chief designer Sidney Camm submitted his proposals to the Directorate of Technical Development (DTD) for the P1040, a high performance medium range fighter powered by a single Rolls Royce Nene engine. The inclusion of a bifurcated tail pipe in the design allowed for a large quantity of fuel to be carried in the fuselage and for the installation of a rocket motor in the tail for the interceptor role. By the end of 1945 the Air Staff had apparently lost interest in the project, while the Naval Staff decided to go ahead

with it as a support fighter which would not need provision for the rocket motor. Pressure was put on the company to redesign the P1040 to take the new Rolls Royce axial flow AJ65 engine (Avon). However, the length of the engine made it impossible to install in the P1040 type fuselage. Consequently the project continued with the Nene engine as the powerplant. In February 1946 an Instruction to Proceed (ITP) was issued to the Hawker Aircraft Company for manufacture of three prototypes and a structural test specimen.

The first prototype, VP401, was an aerodynamic research aircraft without any armament or other military equipment and powered by a single Rolls Royce Nene 1 centrifugal flow turbo-jet engine giving a thrust of 4500lb. The unswept wings were blended into the fuselage at the roots; this local thickening incorporated the twin wing root air intakes, the main undercarriage and the jet pipe exhausts. The fin was integral

Prototype Hawker P 1040, VP401. (B.Ae.)

with the rear fuselage and the tailplane was mounted midway up the fin, well clear of the jet efflux and wing wake. The wing root intakes and divided jet pipe made it possible for the aircraft to carry its total internal fuel capacity of 410 gallons in two fuselage tanks, one in front and the other behind the engine. The unpressurised cockpit was equipped with a Malcolm type ejection seat.

Because of the inadequacy of Hawker's grass airfield at Langley, VP401 was transported to the A&AEE Boscombe Down where it made its maiden flight on 2 September 1947 with W. Humble at the controls. During this flight the pilot encountered a fairly severe engine related airframe vibration. Three days after the first flight the aircraft was transferred to RAE, Farnborough, for the continuation of the test flight programme. Shortly after its arrival at Farnborough the vibration problem was resolved when the rectangular heat shield fairings, which protected the rear fuselage skin from the jet efflux, were replaced by 'pen-nib' fairings which rectified a flow breakaway defect. This new design of fairing became standard on all later aircraft.

Early in 1949, VP401 had been fitted with a 5000lb thrust Nene 101 and had been entered for the 1949 National Air Races at Elmdon. On 30 July 1949 VP401 flown by Sqn. Ldr. Neville Duke won the Kemsley Challenge Trophy race at an average speed of 508 mph. Two days later Sqn. Ldr. T. S. "Wimpy" Wade flew it in the SBAC Trophy race and won at an average speed of 510 mph. Neville Duke and "Wimpy" Wade also shared the Geoffrey de Havilland Memorial Trophy for the fastest lap with an average speed of 563 mph. Shortly after this event the aircraft was converted into the P1072 by the introduction of an Armstrong Siddeley Snarler rocket motor in the rear fuselage.

The second prototype, VP413, flew for the first time on 3 September 1948. This aircraft from the outset was powered by the 5000lb thrust Nene 101 engine and was the first to be fully navalised, meeting the requirement of Specification N7/46, having wing folding, an arrester hook and four 20mm cannons. After tests on the dummy deck at the A&AEE, Boscombe Down, VP413 was flown on to HMS *Illustrious* during May 1949 for its initial carrier trials. These trials were cut short because of problems with the arrester hook installation and the very bad weather conditions. Following modifications to the arrester hook, VP413 returned to the deck of HMS *Illustrious* in October 1949 when 28 landings were made.

The third and last prototype, VP422, flew for the first time on 17 October 1949. This aircraft was more representative of the intended production version, being fitted with a lengthened arrester hook, pick-up points for underwing drop tanks and provision for rocket assisted take-off gear (RATOG). The initial carrier trials were finally brought to a conclusion when VP422 was flown on to HMS *Illustrious* on 15 February 1950. Fourteen landings were made during the day, before the trials were brought to a premature end by bad weather. Both VP413 and VP422 were used for preliminary service evaluation during 1950.

Hawkers received the first production contract on 22 November 1949 for the manufacture of 151 of these new naval fighters, by then named Sea Hawk, to production specification 25/48P. These differed from the prototypes primarily by the introduction of a

Sea Hawk F Mk 1, WF171/168, of No. 806 Squadron taxying on the wooden deck of USS *Antietam* during trials on the ship's angled deck in June 1953. (B.Ae.)

pressurised cockpit and a Martin-Baker ejection seat. There were also several changes of geometry including a 2ft 6in increase of wing span to improve take-off and landing performance, compensating for the weight growth resulting from the introduction of operational equipment.

Production started at the Hawker factory at Kingston-upon-Thames, but with the introduction of "super-priority" status in 1951 for both the Sea Hawk and Hawker's new RAF fighter, the P1067 Hunter, it was evident that Hawker's production facilities were inadequate to meet both programmes. Consequently development and production of the Sea Hawk was transferred, in 1952, to Armstrong Whitworth's factory at Baginton, Coventry which, like Hawker, was a member of the Hawker Siddeley Group and where there was spare production capacity available.

However before production was transferred, 35 Sea Hawk F Mk 1s had been built at Kingston. Many of these early production aircraft were used for trials and service evaluation. These included: WF143, used for stability and control trials by Hawker and the A&AEE; WF144 and WF145, used for carrier trials on HMS *Eagle*; WF147, fitted with powered ailerons which in effect made it the prototype Sea Hawk F Mk 2; and WF148, which was used for cold weather trials in Canada. The seventh production Sea Hawk, WF149, was used for armament trials at the A&AEE, and was unfortunately destroyed in a fatal accident there on 27 June 1953 when the port wing folded on take-off.

The decision to order the Sea Hawk was by no means unanimous as many people involved in the project believed that the Royal Navy should have selected the swept wing "Sea Hawk" based on the experimental Hawker P1052. In January 1947 Hawker Aircraft Ltd had received Specification E38/46 and during March had submitted proposals for project P1052 to the Ministry of Supply. These proposals were accepted, and in May a contract was placed covering the manufacture of two prototypes, VX272 and VX279, and a structural test specimen. The P1052 was in effect the P1040 fitted with swept wings. VX272 flew for the first time on 19 November 1948 and VX279 on 13 April 1949. Although the P1052 was an Air Ministry project the Admiralty did show some interest in it and plans were made for it to carry out carrier trials. Unfortu-

Sea Hawk F Mk 2, WF274, at the Armstrong Whitworth airfield, Bitteswell, April 1954.

nately VX272, which was allocated for the trials, suffered two crash landings and consequently the carrier trials were much delayed. After fitting VX272 with an arrester hook and the long stroke Sea Hawk undercarriage, VX272 landed aboard HMS *Eagle* during May 1952 to commence the trials and although the trials were successful the results were somewhat academic as by then the Sea Hawk was entering service.

Armstrong Whitworth's first production Sea Hawk was an F Mk 1, WF162, which made its first flight on 18 December 1952 in the hands of their chief test pilot Sqn. Ldr. Eric G. Franklin. The production rate rapidly increased, with the 60th and last AWA built Sea Hawk F Mk 1, WF235, being delivered in December 1953. The Sea Hawk F Mk 2 was the next version produced, and this saw the introduction of fully powered aileron control with spring feel and centring. These modifications to the system corrected a lateral control problem which existed on the F Mk 1 at high speed, caused by undamped oscillation of the ailerons. Only 40 F Mk 2s were produced and all had been delivered before the end of March 1954.

The Mk 3 Sea Hawk was a fighter bomber development of the F Mk 2, with the facility to carry two bombs or mines weighing up to 1,000lb each in place of the long range drop tanks that could be carried under the inner wing of the earlier marks of Sea Hawk. Trials with various combinations of these stores were carried out by AWA at its aerodrome at Bitteswell and by the A&AEE at Boscombe Down, using the first FB Mk 3, WF280, which had flown for the first time on 19 March 1954. A total of 116 FB Mk 3s had been produced by 6 October 1954 when the last, WN119, was delivered to the Receipt and Despatch Unit (RDU) at RNAS Stretton.

To meet the requirement for a photographic reconnaissance fighter, an F95 low-level, forward facing camera was mounted in the nose of an 88 gallon drop tank. This installation was carried, in place of the standard tank under the port wing of WM914, for trials which were subsequently continued using the FGA Mk 6 XE369. Although extensive evaluation was undertaken during 1954/5, the system was not adopted for operational service.

The fifth production Sea Hawk FB Mk 3, WF284, was fitted with a set of outer wings modified to allow the carriage of rocket projectiles and bombs, making it the prototype of the definitive ground attack version, the FGA Mk 4. Tests using WF284 carrying 20 x 60lb rocket projectiles under the outer wings commenced on 14 July 1954 and five flights totalling four hours were made. The tests showed that the increased drag seriously affected the aircraft's performance, restricting its role to that of a comparatively short-range, low level attack aircraft of poor performance. Even after the adverse report of WF284's performance, production of the FGA Mk 4 continued with the first production aircraft, WV792, being completed in August 1954. The batch of 97 FGA Mk 4 aircraft was completed during the following seven months.

In an attempt to raise the critical Mach number, Vortex generators were fitted to the tailplane and elevators of WV825, but trials in April 1955 showed that the improvement was only minimal. This aircraft was subsequently fitted with fully powered elevators and electrically operated elevator trim. Test flying commenced on 15 October 1956 and continued until April 1957. Although the system was shown to improve the controls significantly, it was not introduced in production aircraft.

Flight refuelling trials were carried out in January 1957 using the FGA Mk 4 WV840 as the receiver. For these trials refuelling probes were fitted in the front of standard finned drop tanks. Unfortunately the tests, using a Canberra tanker, were not very satisfactory, mainly due to drogue turbulence, which affected longitudinal control during the final approach to the drogue. Presumably because of the difficulty in correcting this problem and the lack of a suitable naval tanker aircraft, the project was soon abandoned.

During 1954 Rolls Royce introduced the up-rated Nene 103 engine giving a thrust of 5,200lb, an increase of 200lb on the Nene 101 used in the previous marks of Sea Hawk. In order to update its aircraft the Admiralty decided to retrofit the Nene 103 in the FB Mk 3s and FGA Mk 4s, re-designating them FB Mk 5s and FGA Mk 6s respectively. Forty-two FB Mk 3s were modified to FB Mk 5 standard and 76 FGA Mk 4s, including WF284 the FB Mk 3 that had been used as the FGA Mk 4 prototype, were converted to FGA Mk 6s. No new FB Mk 5s were produced but a contract was placed for the production of 87 FGA Mk 6s. These aircraft were all delivered between March 1955 and January 1956. The opportunity was taken to introduce the 'Green Salad' navigation system into the FGA Mk 6, making this the only change in the avionics fit during the production of the Sea Hawk.

Towards the end of 1955 the FGA Mk 6 XE456 was

modified to undertake a long range bomber role. This involved the introduction of additional drop tank stations beyond the outboard bomb stations of each wing, together with a fully automatic fuel transfer system. Trials began in January 1956, the aeroplane carrying four standard Bristol finned plastic drop tanks and two 500lb bombs. Unfortunately the outcome was not very satisfactory; take-off performance was very poor and problems were encountered with the fuel transfer system. As a consequence the project was abandoned in July 1956. In the meantime production of the FGA Mk 6 continued until 10 January 1956 when with the delivery of XE490, the 434th production aircraft, to RNAS Abbotsinch, production of the Sea Hawk for the Fleet Air Arm came to an end.

The Sea Hawk entered first line service in March 1953 when No. 806 "Ace of Diamonds" Squadron reformed at RNAS Brawdy with eight Sea Hawk F Mk 1 aircraft. On 15 June 1953 the squadron, led by its Commanding Officer Lt. Cdr. P. C. S. Chilton, took part in the flypast for the Coronation Review of the Fleet at Spithead. Shortly afterwards, on 23 June, No. 806 commenced trials on the USS *Antietam,* testing the fully angled deck, a British concept that had not at that time been introduced on a British carrier. After working up at RNAS Brawdy No. 806 embarked in HMS *Eagle* on 2 February 1954 for a Mediterranean cruise. The introduction of the Sea Hawk on board HMS *Eagle* resulted in the introduction of a new type of crash barrier. Because of its tricycle undercarriage and the rather vulnerable location of the cockpit in the nose of the aircraft, the new barrier had the upper horizontal strand raised to clear the canopy and larger spacings between the vertical strips to clear the fuselage.

By the end of 1953 Nos. 804 and 898 Squadrons had been equipped with Sea Hawk F Mk 1s but neither was to take its aircraft to sea.

Line-up of Sea Hawk FGA Mk 4s of No. 810 Squadron at Lossiemouth in July 1955.

Sea Hawk FB Mk 3s, WM945/103/Z, WM963/104/Z, WM920/107/Z and WM969/111/Z, of No. 898 Squadron flying over HMS *Albion* in September 1954. (RAF Museum)

The Sea Hawk F Mk 2 entered service in 1954, equipping Nos. 802 and 807 Squadrons and also joining No. 806 although apparently not all the squadron's F Mk 1s were replaced by F Mk 2s. After the fighter versions of the Sea Hawk were phased out of front line service in 1954 various training squadrons, including Nos. 738 and 764, continued to use F Mk 1s and F Mk 2s for several years. Early in 1956 the RNVR fighter squadrons started to replace the Sea Furies and

Attackers with Sea Hawk F Mk 1s. In the event however No. 1832 Squadron of the Southern Air Division was the only one to complete the re-equipment before the Air Branch of the RNVR was disbanded in March 1957. In 1956 Cdr.(A) G. McC. Rutherford DSC, RNVR became the first RNVR pilot to deck land a jet when No. 1832 Squadron undertook its annual training in HMS *Bulwark*. The other RNVR squadrons that were programmed to go to the deck on HMS *Bulwark* had their visits cancelled when *Bulwark* was returned to port to prepare for its operations at Suez.

The Sea Hawk FB Mk 3 entered service with No. 806 Squadron in July 1954, quickly followed by Nos. 800, 801, 803, 811, 897 and 898 Squadrons. Aircraft in these units were progressively replaced by later marks of Sea Hawk as they became available, so that by mid 1956 the front line squadrons were equipped mainly with FGA Mk 6s, although a few FB Mk 5s remained.

At the commencement of operation "Musketeer", the Anglo-French landings at Suez during November 1956, a Royal Navy task force which included HM Ships *Albion*, *Bulwark* and *Eagle*, was on station in the East Mediterranean. Operating from these carriers were six squadrons of Sea Hawks in addition to those equipped with Sea Venoms, Wyverns and Skyraiders. Nos. 800, 802 and 810 operated from *Albion*, Nos. 804 and 897 from *Bulwark* and No. 899 from *Eagle*. On 1 November 1956, the start of the British involvement, the Sea Hawks were used for low level attacks on airfields and other military installations and following the landings provided close support for the ground forces. Although two Sea Hawks were shot down and several damaged by ground fire, they had proved to be very very effective in the ground attack role.

Replacement of the Royal Navy's Sea Hawks commenced in 1958 with the arrival on the scene of the Vickers Supermarine Scimitar. However it was not until 15 December 1960 that the last front line Sea Hawk squadron, No. 806, disbanded at RNAS Brawdy after disembarking from HMS *Albion*. The aircraft did however continue to operate on second line duties for some years. The best known of these were probably the all-black Sea Hawks of the Fleet Requirements Unit (FRU) at Hurn. These civilian operated aircraft were equipped with a Harley light mounted in the nose of the port underwing drop tank and were used on radar calibration duties by Royal Navy warships. These last Sea Hawks started to be withdrawn from service in 1966 with the arrival of the Scimitar, but it was not until February 1969 that the last one, XE390, was withdrawn from use. This was however not quite the end of the Sea Hawk in Royal Naval service, as a number were taken on charge by the School of Aircraft Handling at RNAS Culdrose where, along with various other types, they were used to train aircraft handlers in the techniques of manoeuvring aircraft on carrier decks.

The Sea Hawk proved to be one of the most popular aircraft to see service with the Fleet Air Arm. It was a delight to fly and a pleasure to maintain, making it a worthy successor to the Sea Fury. The excellent aerobatic capability of the Sea Hawk resulted in several squadrons forming aerobatic teams, including No. 804 which, with a team of seven under the leadership of Lt. Cdr. Eric M. Brown OBE, DSC, AFC, is believed to have been one of the earliest teams to break away from the usual four aircraft. However, the Royal Navy's most famous aerobatic team must be the five scarlet Sea Hawks of No. 738 Squadron, led by their commanding officer Lt. Cdr. A. J. Leahy DFC, which

Sea Hawk FB Mk 5, WM994, of the Hal Far Station Flight, 1960.
(A. E. Hughes)

provided a superb demonstration of formation aerobatics at the 1957 SBAC Display at Farnborough.

Although attempts to sell the Sea Hawk to Australia and Canada, navies that had up to that time tended to operate equipment common to the Fleet Air Arm, had come to nothing, Armstrong Whitworth Aircraft Company did however succeed in winning a contract to supply 22 Sea Hawks to the Royal Netherlands Naval Air Service, enabling the production line to be reopened. The next customer was the West German Navy, which purchased 64 Sea Hawks for its land based air arm. These differed from the standard Sea Hawk, primarily by the introduction of a larger fin and rudder. The final customer was the Indian Navy, which ordered 14 new build aircraft in addition to 32 reconditioned ex Fleet Air Arm aircraft and later 28 Sea Hawks from the West German Navy. It is interesting to note that the Indian Naval Air Service continued to operate the Sea Hawk in first line service until they were replaced by the British Aerospace Sea Harrier in 1984, almost exactly 30 years after the Sea Hawk first entered service with the Fleet Air Arm.

Sea Hawk FGA Mk 6 of No. 804 Squadron starting up on HMS *Ark Royal*. Two US Navy F3H Demons from USS *Saratoga* are ranged on the port side. (Royal Navy)

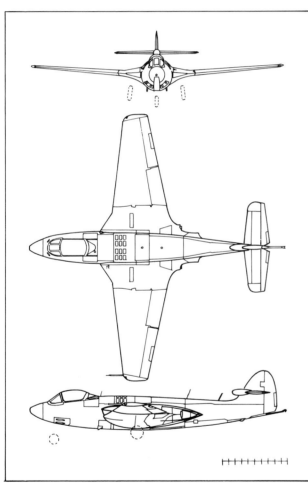

19 Sea Hawk FGA Mk 6

PRODUCTION

P.1040 (Prototype) VP401.
N.7/46 (Prototypes) VP413, VP422.
F Mk 1 WF143 – WF161, WF167 – WF177, WM901 – WM905, WF162 – WF166, WF178 – WF192, WF196 – WF235.
F Mk 2 WF240 – WF279.
FB Mk 3 WF280 – WF289, WF293 – WF303, WM906 – WM945, WM960 – WM999, WN105 – WN199.
FGA Mk 4 WV792 – WV807, WV824 – WV871, WV902 – WV922, XE327 – XE338.
FGA Mk 6 XE339 – XE344, XE362 – XE411, XE435 – XE463, XE490.

CHAPTER TWENTY-ONE
Westland Wyvern

The design of the aircraft which was eventually to be known as the Wyvern began at Yeovil in 1944 under the leadership of the technical director W. E. W. "Teddy" Petter. Designed as a long-range naval strike fighter, the original design located the engine behind the pilot with a propeller shaft passing under the cockpit to the propeller in the nose. This enabled the pilot to be located near the nose of the aircraft, providing him with an excellent forward view for deck landing. Official backing for the project was soon received and Specification N.11/44 was issued to cover it.

The original design however was soon dropped and was replaced by a much more conventional design with the engine located in the nose. The project designated the W.34 by Westland featured a distinctive humped fuselage which provided the pilot with a better forward view than most of the contemporary carrier-borne aircraft; although not as good as would have been the case with the original proposed layout. The new 3500hp 24 cylinder, liquid-cooled, sleeve-valve, Rolls Royce H.46 Eagle engine was selected to power the project. The design was for an orthodox, low wing aircraft with a tail-wheel undercarriage. The wings, which included a double fold to achieve the maximum height limitation for carrier aircraft, were fitted with Youngman flaps

across the centre-section and plain split flaps on the outer wings inboard of the ailerons. The standard armament was four wing mounted 20mm cannons and the aircraft was to have the capability to carry a torpedo under the fuselage and bombs or rocket projectiles under the wings. To keep the propeller diameter to an acceptable size, eight and six bladed contra-rotating propellers were developed by Rotol Ltd and de Havilland Propellers Ltd respectively. A proposed version of the W.34 for the RAF was similar to the RN version, except that it dispensed with the specific naval requirements of wing fold and arrester hook.

In November 1944 six prototypes were ordered and plans were made to follow this with an order for 30 pre-production aircraft: 20 for the Royal Navy and ten for the RAF. At about this time Specification F.13/44 was issued to cover the RAF's requirement for a long range escort fighter specifically for operation in the Far East. However in 1945 it was decided that in future all RAF fighter aircraft were to be jet powered and consequently interest in the Wyvern was abandoned in December 1945. However, as far as the Royal Navy was concerned, experience at that time had shown that propeller driven aircraft were more suited to carrier operation than pure jet types and the Navy was convinced that the Wyvern would satisfy its requirements.

The first prototype Wyvern Mk 1, TS371, was flown

First prototype Wyvern TF Mk 1, TS371, powered by the Rolls Royce Eagle piston engine. (RAF Museum)

First production Wyvern TF Mk 2, VW867, used by Armstrong Siddeley Motors for development of the Python turboprop engine.
(Rolls Royce)

for the first time at the A&AEE Boscombe Down by Westland's chief test pilot Harald Penrose on 12 December 1946. Virtually at the same time as the first flight it had been decided to cancel the Rolls Royce Eagle engine project, although the small initial batch was completed, sufficient only for the prototypes and half of the pre-production batch. This led to ten of the pre-production aircraft being cancelled, and it is believed that in the event only seven were completed. The second prototype, TS375, flew for the first time on 10 September 1947 and a few days later was joined on the test flight programme by the third prototype, TS378.

The first prototype, TS371, was unfortunately destroyed on 15 October 1947 when, after completing an air-to-air photographic sortie for 'Flight' magazine photographer John Yoxall, the propeller bearing failed and the pilot, Sqn. Ldr. Peter Garner, Westland's assistant chief test pilot, was killed attempting to land the aircraft in a small field.

Neither of the first two prototypes was navalised, both lacking folding wings and an arrester hook. They were also unarmed as they were intended to be used for aerodynamic and handling trials only. The third prototype, TS378, was fully navalised, and was fitted with the four wing-mounted cannons. It was also the first Wyvern to be fitted with the de Havilland six-bladed, contra-rotating propeller; the first two having been fitted with the Rotol eight-bladed contra-rotating propeller. The remaining prototypes and all the pre-production aircraft were to full production standard, including armament and full naval equipment.

In April 1948 TS378 was delivered to the A&AEE Boscombe Down for preliminary deck landing assessment and take-off trials. Initial carrier trials commenced on 9 June 1948 when Lt. Cdr. Hickson landed TS378 on HMS *Implacable*. However after only three landings the trials were brought to a premature conclusion when all Wyverns were grounded following the failure of the propeller on TS380, which was the back-up aircraft for the carrier trials. Trials recommenced on 13 July and 15 landings were made on HMS

Implacable that day and a further seven were made with TS380 on the following day. After the application of some minor modifications to the control layout, TS378 was used for further carrier trials, making a total of 100 landings and take-offs on HMS *Illustrious* during May and June 1949.

Following the cancellation of the Eagle engine, a new Specification N.12/45, was issued to cover the development of a turbo-prop Mk 2 version of the Wyvern identified by Westland as the W.35. During February 1946, three prototypes were ordered, one to be powered by the Rolls Royce Clyde engine and two by the Armstrong Siddeley Python, which was a turbo-prop version of the ASX axial-flow gas turbine. The first Wyvern Mk 2 to fly was the Clyde-powered VP120 which flew for the first time on 18 January 1949 with Harald Penrose at the controls on what was to be his shortest test flight, for as the wheels left the ground the cockpit filled with smoke, convincing the pilot that the aircraft was on fire and that he should return the aircraft to the ground as quickly as possible. This he succeeded in achieving after a flight of some three minutes. Fortunately there was no fire, only fumes coming from where fuel was leaking on to an exhaust pipe.

VP120 was delivered to Rolls Royce at Hucknall on 4 July 1949 for engine development. The Clyde engine project was however soon cancelled and VP120, which had only flown for some 50 hours, was transferred to Napier at Luton for the installation of a 4100hp Napier Nomad compound engine. This scheme was abandoned and the aircraft ended its days on crash-barrier trials at the RAE Farnborough.

With the cancellation of the Clyde engine, efforts were concentrated on bringing the Python powered Wyvern TF Mk 2 up to operational standard. As part of the engine development programme, both an Avro Lancaster and Avro Lincoln had Python engines installed in the outboard nacelles by Armstrong Siddeley at Bitteswell. The first Python engined Wyvern VP109 made its initial flight on 22 March 1949 and was followed on 30 August by the second Python Wyvern prototype, VP113. In addition to the change of engine the only other significant difference between the Wyvern Mk 1 and the prototype Mk 2s was the intro-

Formation of Wyvern S Mk 4s of No. 813 Squadron, Ford, in August 1953. (RAF Museum)

duction of the ML ejection seat. This was however changed to Martin Baker ejection seats on all subsequent Wyverns.

With the project committed to the AS Python engine, a batch of 20 Wyvern TF Mk 2s were ordered. The introduction of what was a revolutionary new engine required a lengthy development programme leading to many changes to the powerplant and its control system. It was not until 1952 that the powerplant was considered ready for operational service. During this programme there had been design changes to the airframe, aimed at improving the aircraft's handling. These included an increase of fin area, introduction of dihedral to the tailplane along with small auxiliary fins, and changes to the leading edge section.

Shortly after its first flight the second prototype VP113 was destroyed and the pilot Sqn. Ldr. M. Graves was killed when the aircraft, making an emergency landing at Yeovil following engine failure, ran through the airfield fence and crashed into an adjacent housing estate.

The first of the new Python 2 engines was fitted into the first production TF Mk 2, VW867, and the aircraft was allocated as a replacement for VP113 in the development programme. The poor response to throttle movements was resolved with the Python 2 by running it at constant speed with control being achieved by changing the propeller pitch. The proofing tests of VW867, that were required prior to arrested landings, were carried out at RAE Farnborough before the aircraft proceeded to HMS *Illustrious* for carrier trials. These trials were carried out on 21 June 1950 with Lt. Parker making the first landing. The trials, involving 25 landings, were completed successfully in one day with the flying being shared by two other pilots, Lt. Cdr. Goodhart and Lt. Cdr. Orr Ewing in addition to Lt. Parker. After being exhibited at the SBAC Display at Farnborough in September 1950, VW867 was transferred to Armstrong Siddeley at Bitteswell for development of the constant-speed engine. The aircraft returned to Westland's in February 1954 and was broken-up at Yeovil the following year.

Early in the development programme a two-seat

trainer version of the Wyvern had been proposed and after receiving official support Specification T.12/48 had been issued covering this variant. An order was placed for a single prototype of the W.38 Wyvern T Mk 3 and this aircraft, carrying serial number VZ739, flew for the first time on 11 February 1950. The T Mk 3 was basically a TF Mk 2 with the rear fuselage deepened to accommodate the instructor's cockpit, the two cockpit canopies were joined together by a perspex tunnel, rather like the Sea Fury T Mk 20, and a periscope was provided to improve the instructor's forward view. As no production order was placed, VZ739 was relegated for use by Westland's as a 'hack' until on 3 November 1950 an engine failure forced Sqn. Ldr. D. Colvin to make an emergency wheels-up landing at Seaton in Devon and although the landing was a success the aircraft was written-off after futile attempts to recover it from the marsh in which it had landed.

None of the Wyvern TF Mk 2s was to achieve operational status but a production order for 50 of the next version, the Wyvern TF Mk 4 (later S Mk 4), was placed in 1951. Additionally the last seven of the TF Mk 2 contract were completed as Mk 4s and two of the production TF Mk 2s, VW870 and VW873, were retrospectively converted by Westland's into Mk 4s during 1952. In spite of concerted efforts by Westland, Armstrong Siddeley, the RAE and the A&AEE, the problems with the Wyvern persisted and it was not until late in 1952 that the Wyvern S Mk 4 obtained a limited release for service operation from land bases. Westland immediately set to work bringing a batch of aircraft up to the latest modification standard for delivery to the Fleet Air Arm at RNAS Stretton. There they had service modifications installed before delivery to RNAS Ford for No. 813 Squadron, which had started to replace its Firebrand TF Mk 5s with Wyvern S Mk 4s in May 1953. The last of the Firebrands did not leave Ford until August 1953, by which time the squadron commanded by Lt. Cdr. C. E. Price AFC had a full complement of twelve Wyverns.

To clear the Wyvern S Mk 4 for full operational service, carrier trials were carried out between June and November 1953. The first Wyvern S Mk 4 to land on a carrier deck was VZ750 which was used for trials on the angled deck of USS *Antietam* which was operating in UK waters at the time. During the trials five 'touch-and-go' landings were made with the arrester hook retracted, followed by four arrested landings. The advantage of the angled deck was demonstrated during one attempted landing when the arrester hook bounced over all the wires and the pilot opened up the throttle and took off, going round to land safely at the next attempt. Three aircraft, VZ777, VZ774 and VZ746, were then allocated for carrier trials aboard HMS *Eagle*. On its first landing, VZ777 broke its hook retraction jack and took no further part in the trials. The trials continued with VZ774 making 23 landings, but on the sixth landing of VZ764 the port undercarriage leg broke off and the aircraft was seriously damaged. Unfortunately one of the deck crew was struck by pieces of the shattered propeller and killed and the trials were brought to a premature halt. The carrier trials were finally completed at the end of November 1953 when VZ774 made 48 landings on HMS *Illustrious* using the newly developed mirror sight which, along with the angled deck, was to revolutionise carrier operations.

On 18 May 1954 another test pilot lost his life on the Wyvern programme, when VZ747 on a test flight from Armstrong Siddeley's flight test airfield at Bitteswell crashed at Pailton, near Rugby, killing their chief test pilot Edward Griffiths.

Several second line squadrons operated Wyverns on various trials and development tasks; including 700 and 703 at Ford and 787 at RAF West Raynham. From May 1955 the Wyvern pilot conversion course was undertaken by No. 764 Squadron at Ford but in February 1957 this task was taken over by an independent Wyvern Conversion Unit at Ford who ran it until the unit was disbanded in December 1957.

The Wyverns of No. 813 Squadron then commanded by Lt. Cdr. R. M. Crosley DFC, embarked in HMS *Albion* on 24 September 1954 for a Mediterranean cruise. The Wyverns were soon in trouble, suffering from flame-outs caused by fuel starvation during high 'G' catapult launches. VZ783 was one of the aircraft lost over the front of the ship because of this problem, but the pilot Lt. B. D. Macfarlane made history when after his aircraft had been cut in half by the bow of the ship made his escape by successfully ejecting underwater. After deciding that the 'flame-out' problem made carrier operation too hazardous, the squadron was transferred to RNAS Hal Far in Malta on 16 October 1954 and remained there until the engine problem was resolved. This was eventually achieved by the installation of a recuperator in the fuel system to maintain the fuel flow during the catapult take-off. The squadron re-embarked in HMS *Albion* on 22 March 1955 to return home so that the aircraft could be modified.

The second Wyvern squadron was No. 827 which, commanded by Lt. Cdr. S. J. A. Richardson, reformed on 1 November 1954 at RNAS Ford with nine Wyvern S Mk 4s. The squadron embarked in HMS *Eagle* on 10 May 1955 and was joined by 813 on 4 June 1955 for a summer cruise in the Mediterranean, where joint

Formation of Wyvern S Mk 4s of No. 827 Squadron, Ford, in February 1955. (RAF Museum)

exercises were held with the US Navy. Following exercises in Norwegian waters during September and Home waters in October both squadrons disbanded at Ford in November 1955. Simultaneously with the disbanding of Nos. 813 and 827, two new squadrons No. 830, commanded by Lt. Cdr. C. V. Howard, and No. 831, commanded by Lt. Cdr. S. C. Farquhar, were reformed at Ford.

No. 830 equipped with nine Wyvern S Mk 4s embarked in HMS *Eagle* on 16 April 1956 for exercises in the Mediterranean. The squadron was later involved at the outset of operation "Musketeer" flying 18 sorties against the airfield at Dekheila on the first day, 1 November 1956. During the six days of operations the squadron flew 82 operational sorties: attacking airfields, bridges and vehicles for the loss of two aircraft shot down by ground fire. Fortunately both pilots ejected safely and were picked up by *Eagle*'s SAR helicopter. The squadron returned to the UK on board HMS *Eagle* at the end of 1956 and disbanded at Lee-on-Solent on 5 January 1957.

Wyvern S Mk 4, VZ789/133, of No. 827 Squadron on board HMS *Eagle*. (RAF Museum)

Wyvern S Mk 4 of No. 831 Squadron about to be launched from HMS *Ark Royal*. (F.P.U.)

Although formed at the same time as 830, 831 Squadron was, apart from for two days of deck landing training on HMS *Eagle* in April 1956, to remain land based until it embarked in HMS *Ark Royal* on 9 January 1957 for a cruise in the Mediterranean. After returning to Ford in February 1957 the squadron rejoined *Ark Royal* for a Royal Review in the Moray Firth and subsequently sailed to the USA, there to be involved in joint exercises with the USS *Saratoga* before returning home in July. After two further short spells spent aboard HMS *Ark Royal* in Home waters the squadron disbanded at Ford on 10 December 1957.

Shortly before No. 830 Squadron was disbanded, No. 813 Squadron, commanded by Lt. Cdr. R. W. Halliday AFC, was formed, so that for a few weeks the Fleet Air Arm had three Wyvern squadrons in operation. However 813 was to be the replacement for 830 Squadron in HMS *Eagle* and proved to be the last Wyvern squadron, disbanding on 22 April 1958. From that date the remaining Wyverns rapidly vanished from the scene.

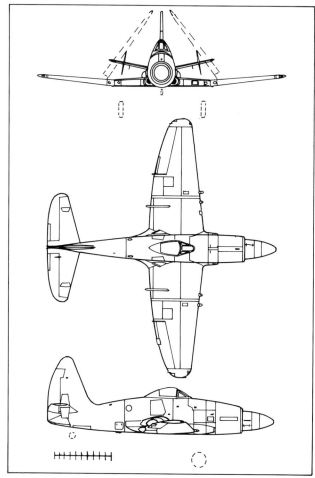

21   Wyvern S Mk 4

PRODUCTION

TF Mk 1 (RR Eagle) TS371 – TS387 (Prototypes).
TF Mk 2 (ASM Python) VP109, VP113 (Prototypes).
TF Mk 2 (RR Clyde) VP120 (Prototype).
TF Mk 2 (ASM Python) VW867 – VW879
T Mk 3 (ASM Python) VZ739 (Prototype).
S Mk 4 (ASM Python)   VW880 – VW886, VZ745 – VZ766, VZ772 – VZ799.

CHAPTER TWENTY-TWO
Hiller HTE

With the rapidly increasing Fleet Air Arm helicopter force there was an urgent requirement for a rugged, reliable helicopter trainer. After giving due consideration to the helicopters available at the time, the Admiralty selected the Hiller HTE-2, which was in large scale production for the US Navy.

The Hiller HTE-2 had started life as a commercial helicopter identified as the Hiller 360, which received its CAA Type Certificate on 14 October 1948, approving it for crop spraying, passenger transport and ambulance duties.

During 1950, the US Marine Corps was using a small number of helicopters in Korea and these were proving so useful for reconnaissance and rescue duties, that there was an urgent demand for a massive increase in helicopter production. At the time, all the helicopter manufacturers in the USA were already heavily involved in the manufacture of military helicopters with the exception of Hiller Helicopters Inc., but it, along with all the other manufacturers, was instructed by the US administration to switch all production to military helicopters. This requirement compelled Hiller to re-design its commercial type 360 to meet a military specification, resulting in the H-23A ordered for the Marines for use as an air ambulance and field evacuation helicopter, and the HTE-1, a two-seat trainer for the US Navy. It was found in service that both these initial military versions were underpowered when using the

178hp Franklin 6V4-178-833 engine. This had been satisfactory in the original commercial 360, but the later military version was some 400lb heavier. Consequently, early in 1951, Hiller commenced a redesign of the military helicopters, incorporating the 200hp Franklin 6V4-200 engine, and the first of these new versions, identified as the H-23B for the Marines and HTE-2 for the Navy, had flown by mid 1951. This more powerful helicopter was a significant improvement over the earlier version and was ordered in large quantities for the Marines and Navy. It was from the large US Navy production contract that 20 aircraft were supplied to the Royal Navy, under the Mutual Defence Aid Programme (MDAP). Deliveries to the UK started towards the end of 1952 and were completed in February 1954.

The Hiller HT Mk 1 entered Fleet Air Arm service in May 1953 when the first aircraft were received at RNAS Gosport by No. 705 Squadron under the command of Lt. Cdr. H. R. Spedding, MBE. The squadron operated as a combined helicoper fleet requirements/basic training/trials unit and, to meet these varied demands, operated Dragonfly and Whirl-wind helicopters alongside the Hillers. Being dual control, the Hiller HT Mk 1s were used almost exclusively for basic helicopter pilot training, and in that role proved to be ideal. It was easy to maintain, relatively easy to fly, reliable, and yet rugged enough to stand the rigours of basic training. Shortly after receiving the Hiller HT Mk 1s, No. 705 was given the honour of leading the Coronation Review Flypast on 15

Hiller HT Mk 1, XB521, being flown by Shorts' test pilot Tom Brooke-Smith at Belfast, January 1955. (Shorts)

CHAPTER TWENTY-THREE
Grumman Avenger

Design work on the Avenger started in 1939 under the designation TBF-1, as a replacement for the US Navy's Douglas TBD-1 Devastator. The TBF-1 was a three-seat, carrier-borne, torpedo-bomber and anti-submarine aircraft capable of carrying a single 22in torpedo or four 500lb bombs in the weapons bay and was to be armed with a 0.5in machine gun mounted in the forward fuselage decking and firing through the propeller arc as well as two rearward facing guns. These were a 0.5in machine gun in an electrically operated dorsal turret mounted aft of the cockpit and a 0.3in machine gun in the rear ventral position. The TBF-1 was of conventional all-metal, stressed-skin construction with a large fuselage providing plenty of space for the aircrew and weapons. Power was provided by a single Wright Cyclone GR-2600-8 engine driving a three bladed propeller. The prototype, XTBF-1, flew for the first time on 7 August 1941, but was destroyed in an accident on 28 November 1941. Fortunately the development programme was not seriously delayed as the second prototype flew on 15 December 1941. Deliveries of the TBF-1, which at about this time had been named the Avenger, started in January 1942 with the production rate rapidly building up to 60 per month by June and achieving an average of 150 per month in 1943. However, because the Hellcat had been given a higher priority, production of the Avenger started to be transferred to the Eastern Aircraft Division of General

Motors Corporation where the aircraft was re-designated the TBM-1. Production of the Avenger finished at Grumman in December 1943 after it had built 2291 and by the end of 1945 when production ended at the Eastern Aircraft Division, 7546 had been built. The final production version of the Avenger was the TBM-3 which was powered by the later Wright Cyclone GR-2600-20 engine.

Avengers started to enter service in the US Navy early in 1942, were in action against the Japanese in June of that year and were to operate with great distinction throughout the rest of the war. They were first supplied to the Fleet Air Arm in 1943 under the Lend-Lease system, and initially the type was officially named Tarpon by the Admiralty, but reverted to Avenger in January 1944. The initial batch received by the Fleet Air Arm were TBF-1s identified by the Fleet Air Arm as the Avenger (Tarpon) TR Mk I, these were followed by TBM-1s which became Avenger TR Mk IIs and finally a batch of TBM-3s which became Avenger TR Mk IIIs. Avengers equipped 15 Fleet Air Arm first line squadrons during the Second World War, primarily in the bombing and strike role, in both the European and Far Eastern theatres of war. With the end of the war the Fleet Air Arm's Avenger squadrons started to disband, and by mid 1946 the Avenger had virtually disappeared from the scene.

With the delays in the development of the anti-submarine Gannet and the increasingly urgent need to replace the obsolescent anti-submarine Barracudas and Fireflies, it was decided that as an interim measure the

Avenger TBM-3E, XB444/065/FD, of No. 703 Squadron, Ford. This aircraft was later converted into an AS Mk 6. (B. J. Lowe)

Avenger AS Mk 4, XB329/064/FD, of No. 703 Squadron, Ford.
(B. J. Lowe)

Avenger should be reintroduced into Fleet Air Arm service. Fortunately at this time the US Navy had started to phase its Avengers out of service and a quantity of these aircraft, TBM-3Es, were allocated to the Fleet Air Arm. The first batch arrived in the UK on 31 March 1953 and by the end of the year a total of 110 had been delivered. A conversion programme was set up at Scottish Aviation Limited at Prestwick to convert the basic American aircaft to the anti-submarine role by the introduction of ASH radar in the AS 4, and the ASV.19A radar in the AS 5, the scanners of both types of radar being carried in pods mounted under the starboard wing. A small number were converted to AS

Avenger AS Mk 5, XB445/364, of No. 815 Squadron, Ford, aboard HMS *Eagle* in 1954. (RAF Museum)

6s; a version which differed by having the dorsal gun turret removed and the cockpit canopy extended aft. However while this conversion programme was under way, No. 815 Squadron replaced its Barracuda IIIs with eight unmodified Avenger TBM-3Es in May 1953, and No. 824 followed shortly afterwards, in July, replacing its Firefly AS Mk 6s. These two squadrons re-equipped in advance of the modified version to give the squadrons some training and familiarisation of the type.

No. 815 Squadron re-equipped with the AS Mk 4s at Culdrose in January 1954 and almost immediately embarked in HMS *Eagle* for a five month cruise in the Mediterranean. No. 824 Squadron re-equipped at about the same time as No. 815 but was to spend all its time shore based, operating from several different Royal Naval Air Stations until its Avengers were replaced by Gannets in February 1955. The only other

Avenger AS Mk 6, XB446/992/CU, of the Culdrose Station Flight. (A. E. Hughes)

first line squadron to be equipped with Avenger AS Mk 4s was No. 820 which replaced its Firefly AS Mk 6s with eight Avenger AS Mk 4s in February 1954 and embarked in HMS *Centaur* in July for the Mediterranean, flying back in February 1955. In March 1955, No. 820 replaced its Avengers with Gannets, by which time all the Avenger AS Mk 4s had been phased out of first line service, leaving just a few in second line service. With the introduction of the Gannet into service only two first line squadrons were to be equipped with Avenger AS Mk 5s and these were: No. 814, which operated the type from March 1954 until November 1955, including some time aboard HMS

Centaur; and No. 815 from July 1954 until October 1955, including operations from HMS *Illustrious* and HMS *Albion*.

Initially only six Avengers were converted to AS Mk 6s, although later a small number of AS Mk 4s and AS Mk 5s were also converted. These were used primarily by No. 831 Squadron, the Fleet Air Arm's Electronic warfare unit which operated four of them with 'A' Flight from May 1958 to June 1959. Orange Harvest was installed in four Avenger AS Mk 5s which were redesignated TS Mk 5s and used by No. 745 Squadron, the radar jamming trials unit, at RNAS Eglinton for tactical evaluation from April 1956 until October 1957.

In November 1955, No. 1830 Squadron of the Scottish Air Division (SAD) of the Royal Naval Volunteer Reserve (RNVR) at Abbotsinch re-equipped with Avenger AS Mk 5s, followed in December by No. 1841 Squadron of the Northern Air Division (NAD) at Stretton and finally by No. 1844 Squadron of the Midland Air Division (MAD) at Bramcote in March 1956. However, the RNVR's use of the type came to a sudden and premature end in March 1957 with the disbanding of the Air Branch of the RNVR.

As the Avengers were phased out of service they were put into storage, mainly at Abbotsinch and although some were subsequently scrapped there, the majority were refurbished and transferred to the Dutch and French Navies during 1957/1958.

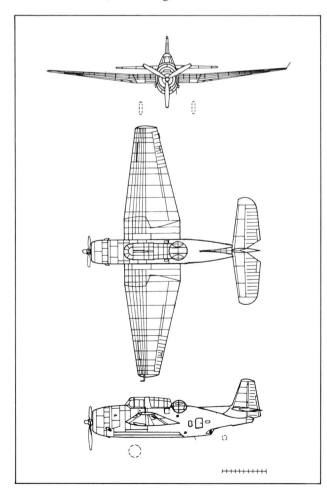

23   Avenger AS Mk 4

PRODUCTION

TBM-3E   XB316, XB318, XB388, XB390, XB442, XB443.

AS Mk 4   XB296, XB301, XB302, XB304, XB306, XB307, XB310, XB313, XB314, XB317, XB319, XB321 – XB330, XB332, XB355, XB356, XB358, XB361, XB362, XB365, XB366, XB438, XB441, XB446.

AS Mk 5   XB297 – XB300, XB303, XB305, XB308, XB309, XB312, XB315, XB320, XB331, XB357, XB359, XB363, XB367 – XB374, XB377 – XB387, XB389, XB391 – XB404, XB437, XB439, XB440, XB445, XB447 – XB449.

AS Mk 6   XB311, XB360, XB364, XB375, XB376, XB444.

CHAPTER TWENTY-FOUR

Boulton Paul Sea Balliol

At the end of the Second World War the Air Ministry issued Specification T.7/45 for an all-purpose, three-seat, advanced trainer to be powered by a turbo-prop engine. Both A. V. Roe and Boulton Paul tendered to the specification and the Air Ministry placed contracts for three prototypes with each company.

The only suitable engines available for the projects were the Armstrong Siddeley Mamba and the Rolls Royce Dart and while Avro decided to use both engine types installing the Mamba in two prototypes and the Dart in the third, Boulton Paul decided to concentrate exclusively on the Mamba. There were however delays with the turbo-prop engines so as an interim measure Boulton Paul installed a Bristol Mercury 30 radial engine in the first prototype of their Type P.108 project which was subsequently named Balliol. This aircraft, VL892, flew for the first time on 30 May 1947, over twelve months earlier than its competitor now named the Avro Type 701 Athena, which had presumably been delayed by the late delivery of the Mamba engine. The remaining Balliol prototypes were fitted with Mamba engines and VL892 was subsequently re-engined with a Mamba.

In 1947 the Air Ministry changed its requirements and issued Specification T.14/47 for a conventional two-seat advanced trainer to be powered by a Rolls Royce Merlin 35 engine, intended primarily as a

Harvard replacement. Both Avro and Boulton Paul revised their previous projects in line with the new specification and orders were placed for four proto-types of the Mk 2 versions of each project. The first Balliol T Mk 2 prototype, VW897, flew for the first time on 10 July 1948 and the first Athena T Mk 2 prototype, VW890, on 1 August 1948. Contracts were soon placed with both companies for 17 pre-production aircraft, but in due course, the last two Athenas were cancelled and the ensuing production contract was placed with Boulton Paul for the Balliol. In all a total of 183 Balliol T Mk 2s were built for the RAF, including 30 built under a sub-contract by Blackburn Aircraft Limited.

The Fleet Air Arm started to take an interest in the Balliol early in 1950 and one of the pre-production T Mk 2s, VR599, was transferred to Royal Navy charge and converted by Boulton Paul into the prototype Sea Balliol T Mk 21. In July 1950 two further Balliol T Mk 2s were loaned to the Royal Navy for evaluation. The first of these, VR598, was initially used for trials at RNAS Arbroath before being navalised, primarily by the introduction of an arrester hook by Boulton Paul. The aircraft was then delivered to the A&AEE for further trials including airfield dummy deck landings (ADDLS), before going back to Boulton Paul in December 1950 to be prepared for return to the RAF. The second of the Balliols loaned to the Royal Navy, VR596, is believed to have been navalised at Boulton

Prototype Sea Balliol T Mk 21, VR599.

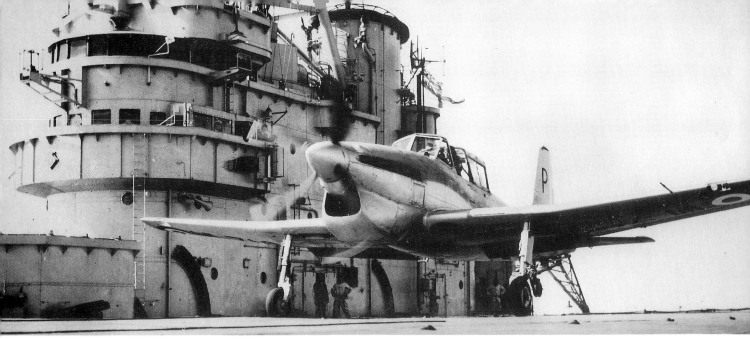

Sea Balliol T Mk 21, WL715/931/P, of HMS *Triumph*'s Ship's Flight, taking off from HMS *Triumph*.

Paul before delivery to RNAS Arbroath in October 1950 where it took over from VR598 as the trials aircraft. In November 1950, this aircraft undertook carrier trials aboard HMS *Illustrious* and was later transferred to No. 703 Squadron, the Service Trials Unit at RNAS Ford, for further trials before being returned to Boulton Paul in July 1951 to be restored to T Mk 2 standard for the RAF.

The prototype Sea Balliol T Mk 21, VR599, flew for the first time on 23 October 1952 and was retained by Boulton Paul for trials until March 1954 when it was delivered to the A&AEE Boscombe Down where it spent the next few years on communications duties at the establishment.

A total of 30 Sea Balliol T Mk 21s were ordered for the Fleet Air Arm and these were delivered between September 1953 and December 1954. The first unit to use the Sea Balliol was the ship's flight of HMS *Triumph* which was formed at RNAS Lee-on-Solent in November 1953 with three Sea Balliols and embarked in HMS *Triumph* on 18 January 1954. This was the only carrier to have Sea Balliols in a ship's flight and operated them until December 1955. In addition to being used by the ship's flight of HMS *Triumph*, Sea Balliols were also used by four Royal Naval Air Stations in their station flights. With its very spacious cockpit the aircraft proved very popular for this communications and general duties role. However events really overtook the Sea Balliol with regard to the trainer role for which it was designed, as the rapid move to jet powered aircraft in the 1950s limited the usefulness of a piston engined trainer. The main user of the Sea Balliol in its training role was the Junior Officers Air Course (JOAC), which gave basic flying training to junior naval officers who were destined for non-flying careers in the Royal Navy. This unit, which used Sea Balliols from October 1957 until July 1959, was usually attached as a flight to one of the second line squadrons and during the period it operated Sea Balliols it spent time attached to Nos. 702, 781 and 727 Squadrons. During 1954 and 1955 Sea Balliols replaced the Harvards used by the RNVR squadrons and subsequently they were operated by a total of four squadrons, some still being on the RNVR strength when it was disbanded in March 1957.

All the Sea Balliols had disappeared from the operational scene by September 1963 with the exception of WL732 which was retained at the A&AEE Boscombe Down for communication duties. When finally withdrawn from use in January 1969 it was allocated to the RAF Museum and is currently displayed at the Cosford Aerospace Museum.

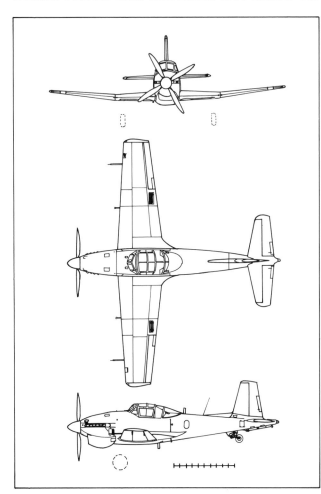

24   Sea Balliol T Mk 21

PRODUCTION

T Mk 21 VR599 (Prototype), WL715 – WL734, WP324 – WP333.

CHAPTER TWENTY-FIVE
De Havilland Sea Venom

When the 4850lb thrust D.H. Ghost engine became available in 1948 the de Havilland Aircraft Company started to look into the possibility of fitting it into the Vampire. The design department decided that in order to obtain maximum benefit from the increased power a much more extensive redesign was necessary, in particular to improve the aerodynamics. The result was a new wing with the leading edge swept back by some 17 degrees and the thickness/chord ratio reduced to 10% from the Vampire's 14%. Additionally the wing was strengthened to carry 75 gallon fuel tanks at the wing tips. In fact the wings were stressed to allow the full range of aerobatic manoeuvres to be carried out with the tip-tanks full.

In the late 1940s it became apparent that an interim fighter-bomber would be required to bridge the gap between the obsolescent Vampires and Meteors then in service and the Hunters and Swifts still at the early stages of development. Consequently the Air Ministry issued Specification F.15/49 which was met by the de Havilland project, initially identified as the Vampire FB Mk 8, but, being significantly different from the DH103 Vampire, subsequently re-identified as the DH112 Venom.

Two Venom prototypes were ordered and to speed up the programme two Vampire FB Mk 5s on the production line were diverted for conversion into Venoms. The first of these, VV612, made its maiden flight at Hatfield on 2 September 1949 to be followed on 23 July 1950 by the second prototype, VV613. Production of the Venom FB Mk 1 for the RAF was soon under way with the first 15 production aircraft being completed at Hatfield during 1951. The company then switched production to their factory at Chester, the first aircraft, WE270, coming off the production line there in June 1952. Production of Venom FB Mk 1s was also sub-contracted to Fairey Aviation and Marshall's of Cambridge.

Once production of the Venom FB Mk 1 was under way, design effort was concentrated on a two-seat, night fighter version. This version, identified as the Venom NF Mk 2 and planned as a replacement for the Vampire NF Mk 10, was similar to the Venom FB Mk 1 but had a larger fuselage nacelle to accommodate the AI (airborne interception) radar in the nose, and side by side seating for the pilot and observer. The first prototype was built as a private venture and flew for the first time at Hatfield on 22 August 1950 carrying the company's class 'B' markings G-5-3, although it was later given the serial number WP227. A total of 90

Formation of Sea Venom FAW Mk 20s of No. 890 Squadron, Yeovilton, led by WM515/206, in 1954. (RAF Museum)

The final production of Sea Venom FAW Mk 22, XG737/438/B, of No. 891 Squadron, HMS *Bulwark*, 1958. (via B. J. Lowe)

No. 890 was absorbed by No. 893 Squadron to bring the squadron up to operational strength.

At the start of the Suez crisis, (Operation Musketeer), there were three squadrons operating Sea Venom FAW Mk 21s in the Eastern Mediterranean. These were No. 809 in HMS *Albion* and Nos. 892 and 893 in HMS *Eagle*. Although being used in their primary interception role, providing defensive cover for the fleet and the allied landings at Port Said, the lack of enemy air activity allowed the Sea Venoms to be used also in their secondary role of ground attack aircraft. Armed with rockets and cannons they were employed to good effect in attacks on enemy targets, including airfields and military vehicles. During the six days of the operation, no Sea Venoms were lost, although a No. 893 Squadron aircraft, WW284/095, was forced to make a wheels-up landing on HMS *Eagle* after being damaged by Egyptian flak.

In January 1957, Sea Venom FAW Mk 22s were delivered to No. 894 Squadron, which had reformed on 14 January at RNAS Merryfield, under the command of Lt. Cdr. P. G. Young. Initially, the squadron operated a mix of Mk 21 and 22s, but during March all the Mk 21s were replaced and the squadron was operating a full complement of twelve Sea Venom FAW Mk 22s. The squadron took its Sea Venoms to sea for the first time on 5 August 1957, when it embarked in HMS *Eagle*. On 20 May 1958 the squadron flew aboard HMS *Eagle* and was to spend the rest of the year in the Mediterranean, returning to the UK in December. The following year, after a short initial Mediterranean cruise

Sea Venom FAW Mk 21, WW186/732, of the Air Direction School, Yeovilton, 1968. (G. A. Jenks)

on HMS *Eagle*, the squadron was predominantly operating from Yeovilton, although detachments were provided to operate from HMS *Victorious*. The squadron joined HMS *Albion* on 1 February 1960 and operated in the Far East until returning to Yeovilton in December 1960, being disbanded there on 17 December. Only two other front line squadrons were to operate Sea Venom FAW Mk 22s. These were No. 891 commanded by Lt. Cdr. J. F. Blunden, which reformed at Merryfield in December 1957 and No. 893, commanded by Lt. Cdr. E. V. H. Manuel, which replaced its Sea Venom FAW Mk 21s with Mk 22s in January 1959. No. 891 was the last Sea Venom squadron to see active service when serving aboard HMS *Centaur* in 1960 it had become involved in Operation 'Damen', carrying out rocket attacks against Yemeni rebel infiltrations in Aden. By the beginning of 1960 all the Sea Venom front line squadrons had disbanded with the exception of the three operating Sea Venom FAW Mk 22s, although they had not much longer to go, with No. 893 disbanding in February 1960, No. 894 in December 1960, and finally No. 891 the following July.

This was, however, not quite the end of front line squadron operation of the Sea Venom, for in June 1957 specially adapted Sea Venoms had been used to equip 'A' Flight of No. 751 Squadron, a Radio Warfare Unit, at RAF Watton. In March 1958, the squadron was re-titled the Electronic Warfare Unit and the modified Sea Venoms were identified as the Mk 21 ECM. The importance of the unit was acknowledged when on 1 May 1958 the squadron achieved first line status and became No. 831 Squadron, comprising 'A' Flight, equipped with four Avenger AS Mk 6s and 'B' Flight with four Sea Venom Mk 21 ECMs. In April 1960 'B' Flight received five Sea Venom Mk 22 ECMs which it operated alongside the Mk 21s until the last of these

was phased out in October 1964. The Mk 22s, however, continued to be operated until the squadron disbanded at RAF Watton on 16 May 1966. The squadron's Sea Venoms spent much of their time on detachment to various stations at home and abroad, including the carriers *Eagle, Victorious* and *Ark Royal*.

Sea Venoms were used by several second line units, including No. 700, the trials and requirement unit which operated them from January 1956 to March 1961. In addition Nos. 736 and 738 Squadrons, the Operational Flying School, used them also from January 1956 to March 1961. A small number were supplied to No. 787 Squadron, the Air Fighting Development Unit at RAF West Raynham, where they were used for the last six months of 1955. The last Fleet Air Arm squadron to use the Sea Venom was No. 750, the Observer School, which received its first four Sea Venom FAW Mk 21s during July 1960, when the squadron was based at Hal Far in Malta, enabling the student observers, following initial training in the squadron's Sea Princes, to be trained as observers for the Fleet Air Arm's high performance Sea Vixen and Buccaneer aircraft. No. 750 Squadron's Sea Venom FAW Mk 21s were replaced by FAW Mk 22s during the second half of 1961 and at the same time, the complement was increased to five aircraft. After moving to Lossiemouth on 23 June 1965, the squadron continued to operate the aircraft until 24 March 1970, when the Sea Venom element of the squadron was disbanded.

The only other unit to operate Sea Venoms was the civilian operated Air Direction Training Unit, operated by Airwork Ltd. at St. David's, a satellite of RNAS

Brawdy. The unit received a number of Sea Venom FAW Mk 20s in October 1955, to replace its Sea Hornets and to operate alongside the squadron's Mosquito T Mk 3, Meteor T Mk 7s and Attacker F Mk 1s and FB Mk 2s. The unit took delivery of its first Sea Venom FAW Mk 21s in February 1957, but the last of the FAW Mk 20s was not phased out of service until June 1959, by which time the unit had moved to Brawdy. Sea Venom FAW Mk 22s started to replace the FAW Mk 21s in January 1961 and the replacement programme was completed in April. The FAW Mk 22s continued to operate with the unit until October 1970, when the last official Sea Venom flight was made with XG683 being delivered from Yeovilton to Culdrose.

In addition to the aircraft supplied to the Royal Navy, de Havilland's also succeeded in selling the Sea Venom to Australia. After initially forming a squadron (No. 808) in the UK with Sea Venom FAW Mk 20s on loan from the Fleet Air Arm, the Royal Australian Navy placed an order for 39 Sea Venom FAW Mk 53s. This version was basically the same as the Fleet Air Arm's FAW Mk 21, but with different equipment to meet the RAN requirements.

The Sea Venom was also selected for the French Navy, and the type identified as the Aquilon was built under licence by Sud Aviation. Following the manufacture of four prototypes identified as the Aquilon 20, a total of 25 of the Aquilon 201, the first production version, was built similar to the Fleet Air Arm's FAW Mk 20. The next version, the Aquilon 202, was significantly different with the installation of ejection seats, and a rearward sliding cockpit canopy. This was followed by the Aquilon 203, a single-seat, all-weather fighter, equipped with APQ94 radar and a normal single-seat fighter canopy. Finally, a number of the Aquilon 201s were converted into the Aquilon 204, a two-seat, all-weather fighter trainer. In all, a total of 25 Aquilon 202s and 40 Aquilon 203s were built. The Aquilons were to give good service with the French Navy, including operations during the Algerian campaign, until they were eventually replaced by the Etendard in 1962.

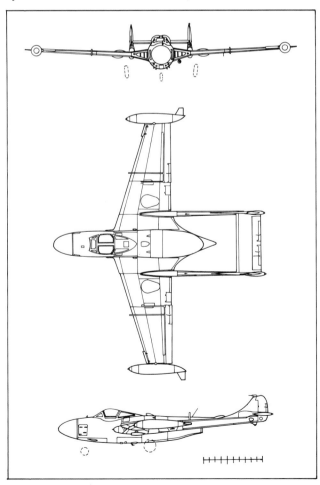

25 Sea Venom FAW Mk 21

PRODUCTION

FAW Mk 20 WK376, WK379, WK385 (Prototypes), WM500 – WM523, WM542 – WM567.

FAW Mk 21 XA539 (Prototype), WM568 – WM577, WW137 – WW154, WW186 – WW225, WW261 – WW298, XG606 – XG638, XG653 – XG680.

FAW Mk 22 XG681 – XG702, XG721 – XG737.

CHAPTER TWENTY-SIX

Fairey Gannet

World War Two had emphasised the Royal Navy's requirement for effective anti-submarine patrol aircraft. To cover this requirement, Specification No. G.R.17/45 was issued for a two-seat, anti-submarine and strike aircraft. The Fairey Aviation Company, which had already carried out some initial design work in anticipation of the specification, immediately set to work refining the design of their Type Q project for submission to the Admiralty. The Type Q was initially intended to be powered by the Rolls Royce Tweed double propeller turbine engine, but when this engine project was abandoned because of pressure of work at Rolls Royce, Armstrong Siddeley was approached to develop a powerplant using two Mamba propeller turbine engines operating through a common gear box.

Armstrong Siddeley was successful in meeting the requirement for the powerplant with the 2950ehp Double Mamba ASMD 1. This was, basically, two

Mambas mounted side-by-side, linked to a common gear box and driving contra-rotating, co-axial propellers. With each of the propellers being driven by half of the powerplant it made it possible for either half of the powerplant to be shut down and its corresponding propeller feathered enabling the aircraft to operate on the other half only to maximise cruising economy. The system gave most of the advantages of twin-engined operation without the usual assymetrical control problems associated with the loss of an engine.

The Fairey Type Q designed by a team at Fairey's factory at Hayes under the leadership of the chief designer H. E. Chaplin, was a mid-wing monoplane with an inverted gull wing, a retractable tricycle undercarriage and a single fin and rudder. The pilot's cockpit was located close to the nose, giving an excellent forward view, ideal for carrier operations. The observer was located in a cockpit immediately aft of the pilot and just above the wing leading edge. The deep fuselage allowed for a roomy weapons bay, and in addition stores/weapons could be carried under the wings. The radar scanner was housed in a retractable

First prototype Fairey 17, VR546, making the first landing on a carrier by a turbo-prop aircraft on 19 June 1950, during carrier trials aboard HMS *Illustrious*. (Westland)

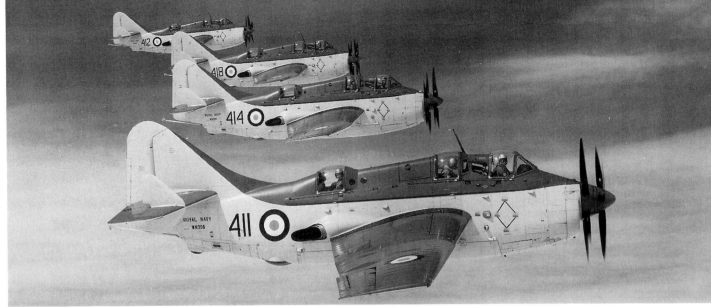

Formation of Gannet AS Mk 1s, led by WN396/411, of No. 824 Squadron, Eglinton, April 1955. (RAF Museum)

radome located in the rear fuselage, aft of the weapons bay. The aircraft was fitted with a sting-type arrester hook and catapult pick-up points for carrier operations. The wings were folded hydraulically, with the inner wings folding up and the outer wings simultaneously folding down.

A contract for two prototypes of the Fairey Type Q was placed on 12 August 1946, and as a precaution against failure of the project, a contract was simultaneously placed for prototypes of its competitor, the Blackburn Y.B.1. The first prototype VR546, by then known as the Fairey 17, was completed in mid 1949 and transported by road to Aldermaston where Fairey's chief test pilot, Group Captain R. G. Slade, commenced taxying trials on 11 September 1949, with the maiden flight taking place on 19 September. Early flights showed that VR546 suffered from serious stability problems with an apparently inadequate elevator control making the aircraft extremely difficult to fly.

On 25 November 1949, shortly after the aircraft had been flown to Fairey's test flight base at White Waltham, these control problems were the cause of a heavy landing, which resulted in the collapse of the nose wheel and ensuing damage to the propellers and airframe. While repairs were being carried out the opportunity was taken to introduce modifications to resolve the control problems. However when flying was resumed on 1 March 1950, the aircraft was still not considered satisfactory and further changes had to be made before VR546 was considered ready for trials at the A&AEE, Boscombe Down in May 1950. Unfortunately, shortly after arrival at the A&AEE, the starboard section of the engine became unserviceable and the aircraft was flown back to White Waltham by Group Captain Slade, using only the port section. After repair, VR546 returned to the A&AEE, where a programme of airfield dummy deck landings were carried out before the aircraft was flown out by Lt. Cdr. G. R. Callingham on 19 June 1950 to make its first landing on HMS *Illustrious* – the first landing on a carrier deck by a propeller-turbine aircraft. By the end of the day, two Fleet Air Arm pilots, Lt. Cdr. Callingham and Lt. R. Reynolds, and the Fairey test pilot, P. Twiss, had made a total of 25 landings.

By the time the second prototype, VR557, flew for the first time, on 6 July 1950, the Admiralty changed the requirement causing a major re-design of the Fairey 17. The most conspicuous change was the introduction of a second observer, located in a cockpit aft of the wing. Further changes included the extension of the weapons bay and re-location of the retractable radome further aft. To meet this revised specification, a third prototype was ordered in June 1949. However, in the meantime, VR546 was modified to act as the aerodynamic prototype of the revised Fairey 17 with the introduction of a mock-up of the new rear cockpit canopy and the radome fitted in the revised location. On its return to the flight development programme, in March 1951, it was discovered that the rear cockpit was having an adverse effect on the airflow over the tail surfaces and to resolve this problem auxiliary fins were fitted to the tailplane during May 1951.

The third prototype, WE488, the first to be built to the revised specification, made its maiden flight on 10 May 1951, by which time the type had been officially named Gannet. With the Fleet Air Arm in desperate need of a modern anti-submarine aircraft, the Gannet was ordered into super priority production with a contract being placed for 100 AS Mk 1s on 14 March 1951. Production of the Gannet utilised Fairey's newly developed envelope tooling which basically involved assembling the structure to the skin producing an accurate external shape and excellent interchangeability.

Meanwhile the three prototypes continued the flight development programme which included over 250 carrier landings and take-offs. The third prototype, WE488, crash landed at Turnhouse on 9 October 1953. It was so badly damaged that it was considered uneconomic to repair and was relegated to being a ground instructional aircraft at RNAS Arbroath. On completion of its flying programme the first prototype, VR546, was used for barrier trials at Bedford before being dismantled and transported to the Royal Naval Air Yard (RNAY) at Donibristle. There it was stripped of usable spares and the remains disposed of for scrap. The second prototype, VR557, was scrapped at about the same time having been used for hot weather trials in Malta during 1953 and then for engine development by Armstrong Siddeley at Bitteswell and, like the first prototype, ended its useful life carrying out barrier trials at Bedford.

A combination of changes in the requirements and development problems had delayed the entry of the Gannet into service and, as a consequence, 110 Grumman Avengers were obtained from the USA through the Mutual Defence Aid Programme as an interim measure to bridge the gap between the Firefly

Gannet AS Mk 4, XA435/280/J, of No. 814 Squadron, HMS *Eagle*, October 1957.

and the Gannets. The first of these Avengers entered service in May 1953, just one month before the first production Gannet AS Mk 1, WN339, made its first flight from RAF Northolt. By this time several refinements had been introduced on the Gannet including the repositioning of the main undercarriage, a revised nosewheel leg and a means of interconnection between the flaps and the tailplane incidence mechanism which would automatically adjust the trim when the flaps were operated. On 13 June 1953, just five days after its first flight, WN339 took part in the Coronation Naval Review Flypast from Lee-on-Solent. All the early production Gannets were allocated to development programmes at Fairey's White Waltham, and Armstrong Siddeley's at Bitteswell.

The sixth production aircraft, WN334, was the first Gannet to see service with the Fleet Air Arm, being issued to No. 703 Squadron at Ford on 7 January 1954 but only being retained by the squadron for one month before returning to Fairey's for preparation for cold weather trials in Canada.

No. 703X Flight commanded by Lt. Cdr. F. E. Cowtan was formed at RNAS Ford on 15 March 1954 as the Gannet AS Mk 1 Intensive Flying Trials Unit and received its first four Gannets at a formal handing-over ceremony at Ford on 5 April 1954. The programme, which included carrier and tropical trials, was interrupted when all service Gannets were grounded for two months while modifications were made to the propeller control system to cure an engine compressor stall problem. During the trials one aircraft, WN348, was lost in the sea when the engine failed on take off from the light fleet carrier HMS *Albion* on 25 August 1954 and another, WN364, was damaged when it made a wheels-up landing following hydraulic failure.

With the introduction of turbine powered aircraft Fleet Air Arm pilots were having to adapt to new flying techniques and to assist with this conversion it was decided to develop a dual control trainer version of the Gannet. To meet this requirement an early production AS Mk 1, WN365, was converted on the production line at Hayes into the prototype of the T Mk 2. Duplicated controls were fitted in the two forward cockpits with a periscope for the instructor who was located in the second of these; all the radar equipment including the retractable radome was deleted and the third cockpit at the rear was modified to accommodate either a radio operator or up to two passengers. The prototype WN365 flew for the first time at RAF Northolt on 16 August 1954. By the following March production T Mk 2s were beginning to enter service

with No. 737 Squadron, the Naval Anti-Submarine School at RNAS Eglinton.

In 1954 production of Gannet AS Mk 1s started at Fairey's Heaton Chapel factory in Stockport and the first Stockport built Gannet, WN370, made its maiden flight from Ringway – now Manchester International Airport – on 5 October 1954 in the hands of David Masters, the company's senior test pilot.

The first operational squadron to receive the AS Mk 1s was No. 826, commanded by Lt. Cdr. G. F. Birch, which replaced its Firefly AS Mk 6s with eight Gannet AS Mk 1s in January 1955. A detachment of four aircraft from this squadron embarked in HMS *Eagle* in May 1955 for one week's deck landing practice and the following month the whole squadron embarked in HMS *Eagle,* to spend the summer operating in the Mediterranean. The second Gannet squadron was No. 824, commanded by Lt. Cdr. J. H. Honywill at RNAS Eglinton which replaced its Grumman Avenger AS Mk 4s with eight Gannet AS Mk 1s during February 1955. This squadron carried out its deck landing practice on HMS *Bulwark* in June 1955, before joining HMS *Ark Royal* on 5 October 1955 for six months in the Mediterranean. No. 820 Squadron, commanded by Lt. Cdr. A. D. Cassidi at RNAS Eglinton, was the third squadron to re-equip with Gannet AS Mk 1s, replacing its Avenger AS Mk 4s on 7 March 1955. After embarking in HMS *Bulwark* in September 1955 for an exercise, the squadron joined HMS *Centaur* in January 1956, becoming the first Gannet squadron to operate in the Far East. Two further squadrons, Nos. 825 and 812, re-equipped with Gannet AS Mk 1s before the end of 1955, followed by the final two squadrons, Nos. 815 and 847, which received their aircraft early in 1956.

The Royal Naval Volunteer Reserve anti-submarine squadrons started to re-equip in February 1956 when 1840/1842 at RNAS Ford received eleven Gannet AS Mk 1s and a Gannet T Mk 2. Plans to re-equip several other RNVR anti-submarine squadrons were overtaken in 1957 by the decision to disband the Air Branch of the RNVR.

Meanwhile continuing development of the Double Mamba had resulted in the next version, the 3035ehp Double Mamba 101 (ASMD3). This engine was installed in the Gannet AS Mk 1, WN372, making it the prototype of the AS Mk 4. This flew for the first time with the new engine on 12 March 1956, only a few weeks before the first production Gannet AS Mk 4, XA412, made its maiden flight.

In line with this development, the Gannet T Mk 2 was also updated by the introduction of the improved engine to become the T Mk 5. The final eight T Mk 2s on the production line at Hayes were completed as T

Mk 5s and the first of these, XG882, flew on 1 March 1957.

No. 824 Squadron commanded by Lt. Cdr. L. D. Urry at RNAS Culdrose introduced the Gannet AS Mk 4 into service when it replaced its existing AS Mk 1s in October 1956. The squadron operated from HMS *Ark Royal* and HMS *Albion* before disbanding on 1 November 1957. Three squadrons, Nos. 814, 825 and 815, re-equipped with AS Mk 4s during 1957, followed later by No. 847 and finally No. 810, commanded by Lt. Cdr. A. Mck. Sinclair, which reformed with AS Mk 4s on 20 April 1959. This latter squadron proved to be the Fleet Air Arm's final fixed wing anti-submarine squadron, for when it disbanded on 12 July 1960 the role had been taken over by the Westland Whirlwind HAS Mk 7. This was, however, not quite the end of the Fleet Air Arm's Gannet AS Mk 4s, for during 1960/61 six were converted to COD Mk 4 standard and used for Courier Onboard Delivery (COD) duties by No. 849 Squadron from September 1961 and a further seven were converted into Gannet ECM Mk 6s for use by No. 831 Squadron in the Electronic Countermeasures role from February 1961.

The Fairey Aviation Company was also successful in winning export orders for the Gannet. The first was for 33 AS Mk 1s and three T Mk 2s for the Royal Australian Navy. These aircraft, built in conjunction with the Fleet Air Arm order, were delivered in 1955 and continued in service until August 1967. The Gannet AS Mk 4 was selected as the initial equipment of The German Federal Republic's Naval (Kriegsmarine) Air Arm anti-submarine unit M.F.G.1 and a total of 15 AS Mk 4s and one T Mk 5 were diverted from the Royal Navy's production order and delivered during 1958. The final export order was from Indonesia, which ordered 18 reconditioned Gannets, including anti-submarine and trainer versions, in January 1959. For this order Fairey purchased 20 Gannet AS Mk 1s and two Gannet T Mk 2s from the Royal Navy to be refurbished and converted into AS Mk 4s and T Mk 5s. A total of 18 AS Mk 4s and two T Mk 5s were delivered to Indonesia along with two airframes for ground training. In addition one aircraft was damaged in a belly landing during training at White Waltham and apparently scrapped and replaced to make up the order quantity before completion of the delivery.

At an early stage of development of the Gannet consideration was being given to an Airborne Early Warning version as a replacement for the piston

engined Douglas Skyraider which had been supplied by the USA under the Mutual Defence Aid Programme. The designation Gannet AEW Mk 3 was allocated, even though the design and development of the AEW version was apparently given a relatively low priority. To meet the AEW requirement a major redesign of the fuselage was necessary, involving the deletion of the two rear cockpits, the weapons bay and the retractable radar radome. A large radome, to house the scanner for the AN/APS-20F radar, was located under the fuselage aft of the nosewheel bay. This necessitated an increase in the area of the fin, and to provide adequate ground clearance, a longer undercarriage had also to be introduced. The pilot's cockpit and canopy remained virtually identical to that of the anti-submarine Gannets, but in the AEW version the two radar operators were accommodated in a cabin located in the fuselage, and accessed by way of entrance doors located just above the wing trailing edges. The engine used was the later 3875ehp Double Mamba 112 (ASMD8), and flight trials with this engine installed in the Gannet AS Mk 1, WN345, had been carried out by Armstrong Siddeley at Bitteswell during 1956.

An order was placed with Fairey for the manufacture of 44 Gannet AEW Mk 3s including the prototype. Work started at Hayes and the prototype, XJ440, made its first flight on 20 August 1958 from Northolt with chief test pilot Peter Twiss at the controls. Not fitted with any of the radar systems, XJ440 was used purely as an aerodynamic prototype and after accumulating some eleven hours flying time, made its first public appearance at the SBAC show, Farnborough, during September 1958. The aircraft then went on to carry out catapult and arresting trials at RAE Bedford during October, and on 18 November 1958 started carrier trials on HMS *Centaur*. During 1959 XJ440 was used for autopilot trials and by the end of 1959 was transferred to Bristol Siddeley at Filton for engine development trials. XJ440 crashed while on final approach to Filton on 26 April 1960 and was written off.

Production aircraft started coming off the line in November 1958, the first, XL449, flying on 2 December 1958 followed by the second, XL450, on 31 January 1959. During May 1959 the first three production aircraft were used for carrier trials aboard HMS *Victorious* and in October 1959 further carrier trials were carried out on HMS *Victorious* using XL452. On 17 August 1959 No. 700G Flight, the Gannet AEW Mk 3 Intensive Flying Trials Unit commanded by Lt. Cdr. W. Hawley, was formed at RNAS Culdrose. By the time it disbanded on 1 February 1960 the flight's three

Gannet COD Mk 4, XA454, used by Flag Officer Aircraft Carriers (FOAC) HMS *Victorious*.

CHAPTER TWENTY-SEVEN
Blackburn B-54/B-88

Designed to meet Specification G.R.17/45 for a carrier-borne, anti-submarine aircraft, Blackburn's B-54 project, identified under the SBAC designation system as the Y.A.5, was in direct competition with the Fairey 17, later to be named the Gannet. Blackburn's initial ideas were based on a development of the B-48 (Firecrest) which in its turn was a development of the Firebrand. However at a very early stage of design it was realised that a totally new concept was required to meet fully the requirement. The new Napier Double Naiad turbine engine driving a contra-rotating propeller was selected as the power unit. The airframe was of all metal, semi-monocoque construction, with a deep section fuselage to accommodate a capacious weapons bay, while the inverted gull wing, similar to the B-48, allowed the use of a relatively short main undercarriage. The crew of two were accommodated in the

top of the fuselage, above the wing centre-section, under individual sliding bubble canopies giving them the excellent view necessary for satisfactory deck landing. The radar scanner was housed in a retractable radome located in the rear fuselage, just aft of the weapons bay.

In 1949 three prototypes were ordered, and while they were under construction, the development of the Napier Double Naiad engine was abandoned. To reduce any resulting programme delay to a minimum, work was put in hand to install a 2,000 hp Rolls Royce Griffon 56 piston engine driving a 13ft, six-bladed, contra-rotating propeller into the first prototype, WB781. With the installation of the Griffon engine, the Y.A.5 was redesignated the Y.A.7. Peter G. Lawrence flew WB781 on its maiden flight at Brough on 20 September 1949, just one day after the first flight of its competitor the Fairey 17. After trials at the A&AEE Boscombe Down, WB791 was flown to HMS *Illustrious* on 8 February 1950 for initial carrier trials.

The Double Mamba powered B-88 YB1 prototype WB797 formating with the Griffon powered B-54 YA7 prototype WB781. (B.Ae.)

Before the second prototype was completed the naval requirements were changed, increasing the crew to three by the introduction of an additional observer. The second prototype, WB788, was therefore modified by extending the rear cockpit to accommodate the two observers sitting facing each other. Although like the first prototype, the second was also Griffon powered, it differed in some details with an increase in the sweepback of the wing leading edge, a taller fin and the introduction of a narrow, mass balanced, rudder. This aircraft, WB788, designated the Y.A.8, flew for the first time on 3 May 1950 and made its first deck landings on HMS *Illustrious* the following month.

It was however not until the third prototype that Blackburn's was able to provide an aircraft that was fully representative of their project to meet G.R.17/45. This prototype, WB797, the Blackburn B-88, designated the Y.B.1, differed from the previous two prototypes in being fitted, like the Fairey Gannet, with an Armstrong Siddeley Double Mamba turbine engine driving a six-bladed, contra-rotating propeller. Peter

Lawrence flew WB797 on its first flight at Brough on 19 July 1950. After appearing at an air display at Lee-on-Solent during August 1950, WB797 was also demonstrated by Lawrence at the SBAC Display at Farnborough in September. However by this time it would appear that the Admiralty had already selected the Fairey Gannet for the Fleet Air Arm and the development of the Y.B.1 was only continued at a low priority, to offer an alternative in case any serious problems developed with the Gannet.

By 1951, when super-priority production of the Gannet was under way, the three Blackburn prototypes were allocated for research and development purposes. WB781 and WB788 were delivered to the RAE, Farnborough, where they were subsequently scrapped in 1957 and 1956 respectively. The third prototype, WB979, was delivered to Armstrong Siddeley at Bitteswell in 1951 where it was used by the company for Double Mamba development work, being finally scrapped in July 1955.

PRODUCTION – Prototypes only

B-54 (Y.A.7) WB781.
B-54 (Y.A.8) WB788.
B-88 (Y.B.1) WB797.

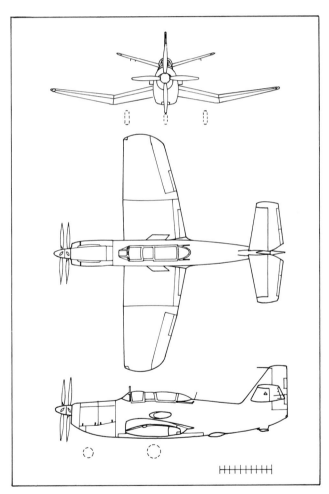

27 Blackburn B-88

CHAPTER TWENTY-EIGHT

Short Seamew

Shortly after World War Two the rapid expansion of the Soviet submarine fleet was causing considerable concern within NATO and it was becoming evident that the only way to meet this increasing threat would be by increasing NATO's anti-submarine force. Unfortunately at the same time the NATO countries were having to operate with limited defence budgets, which made it impossible to provide adequate quantities of very advanced aircraft that were needed to replace those which had been in service since the war. There was however an opinion developing within the Admiralty and certain of the aircraft companies that naval aircraft had become much too complex, heavy and expensive, and that it should be possible to develop simpler aircraft that could be purchased in large quantities. In addition, the relatively low cost of buying and operating such an aircraft would make it suitable for some of the smaller less wealthy members of NATO. These arguments were particularly appropriate in regard to anti-submarine aircraft, because if war was to break out there would be an immediate need for large numbers of aircraft to protect the fleet from the submarine threat.

The Admiralty, convinced by the arguments, issued Specification No. M.123 in 1951 for a simple, light-weight anti-submarine aircraft that could be operated from any of the Royal Navy's aircraft carriers in all but the very worst weather conditions. Rather fortuitously, Rear Admiral Matthew Slattery, who, when serving in

the Royal Navy, had been one of the main advocates of the simpler, cheaper aircraft, had been appointed Managing Director (Technical) by Short Brothers and Harland Ltd only shortly before the specification was issued. Consequently he was keen that Short's should win the contract and immediately set the design team, led by the chief designer David Keith-Lucas, to work preparing the design of an aircraft to meet the specification.

The resulting design was for a fairly basic, single engined, monoplane aircraft of a rather unusual appearance, identified as the Short SB6. The crew of two – pilot and observer – were located in tandem, high above the fuselage nose to give the pilot an excellent forward view, ideal for deck landing. For simplicity the aircraft had a long-stroke, fixed tailwheel type undercarriage. Plans for a more suitable tricycle undercarriage had to be abandoned as the proposed nosewheel obstructed the radar scanner housed in a radome under the nose. The Armstrong Siddeley Mamba turboprop engine was selected in preference to the well proved Rolls Royce Merlin piston engine as it had become Royal Navy policy to phase out the carriage of large quantities of high octane fuel on board ships. In addition it was considered that a turboprop engine was less likely to interfere with the ASV radar and, as it would run with less vibration than a piston engine, would be much more comfortable for the crew.

Of the various designs prepared to Specification M.123 and submitted to the Admiralty the Short SB6, later named Seamew, was selected and a contract for

Second prototype Seamew AS Mk 1, XA213, landing on board HMS *Bulwark* during carrier trials in July 1955.

Seamew AS Mk 1, XA213, on the lift aboard HMS *Bulwark*, July 1955. (Shorts)

two prototypes, XA209 and XA213, was placed in April 1952. Work progressed rapidly and the first prototype, XA209, was flown by the chief test pilot W. J. Runciman on 23 August 1953 at Sydenham. Unfortunately the aircraft was damaged on landing from the first flight but was repaired in time to appear at the SBAC Show at Farnborough the following month. This aircraft was not fitted with any of the operational equipment and consequently was used for aerodynamic development and although it was not fitted with radar, the radome was installed under the nose early in the flying programme. From the outset the Seamew proved to be a very difficult aeroplane to fly, with very unsatisfactory flying controls. To resolve the problems a series of modifications were embodied on the prototype, including the introduction of fixed leading edge slats, slots into the outboard ends of the flaps, standard ailerons to replace the original Frise type and inverted slats under the tailplane roots.

The second prototype, XA213, was the first fully equipped Seamew and joined the flying programme early in 1954. Both prototypes took part in the 1954 SBAC Show, giving a demonstration of the range of speeds the aircraft was capable of achieving when XA209 flew past at a speed close to the stall to be overtaken in front of the crowd by XA213 flying at maximum speed. XA213 was allotted for carrier trials and made its first landing on board HMS *Bulwark* on 12 July 1955. Following a second period on HMS *Bulwark* the trials were completed by December 1955.

Early in the development programme the RAF started to take an interest in the Seamew as a possible type for Coastal Command. It was seen as a supplement to their long range maritime reconnaissance Lancaster and Shackleton aircraft by taking over the shorter range patrols. The result was the Seamew MR Mk 2 which only differed from the Royal Navy's AS Mk 1 by deletion of the equipment necessary for carrier operation and the introduction of larger main wheels fitted with low pressure tyres more suited to land operations. Although the folding wing was retained, the powered folding mechanism was deleted, which meant that the wings had to be folded manually.

In February 1955 a production contract was placed for 60 Seamews, believed to cover equal quantities for the RAF and the Royal Navy, although the latter was given priority with the deliveries. Seamews were delivered to No. 700 Squadron, the Intensive Flying Trials Unit at RNAS Lossiemouth in November 1956 and two aircraft were involved in intensive carrier trials on board HMS *Warrior* making some 200 take-offs and landings, including catapult launches. By this time the RAF appeared to have completely lost interest in the project and its batch of Seamews was cancelled after only four MR Mk 2s had been built. Three of these Mk 2s were subsequently converted into AS Mk 1s and the fourth, XE175, flown by W. J. Runciman, had an unsuccessful sales tour of Italy, Yugoslavia and Germany during the Spring of 1956. After its return to the UK, XE175 was destroyed and W. J. Runciman killed, when it crashed giving an aerobatic display at Sydenham on 9 June 1956.

During 1956 the Admiralty had decided that the Avengers being used by the RNVR anti-submarine squadrons were to be replaced by Seamews. However, after only seven Seamews had been delivered to the Royal Navy, the Air Branch of the RNVR was disbanded in March 1957 as part of the defence cuts, which also resulted in the cancellation of the Seamew production contract. The seven aircraft that had been delivered to the Royal Navy were put in store at RNAS Lossiemouth and scrapped some years later. Also scrapped were ten of the eleven completed aircraft held by Short's at Sydenham awaiting delivery, the exception being one that was used for a while in the apprentice training school.

Following the failure of the Seamew, no further attempts were made to develop a lightweight operational aircraft for the Royal Navy.

PRODUCTION
AS Mk 1 XA209, XA213 (Prototypes), XE169 – XE186, XE205 – XE216.

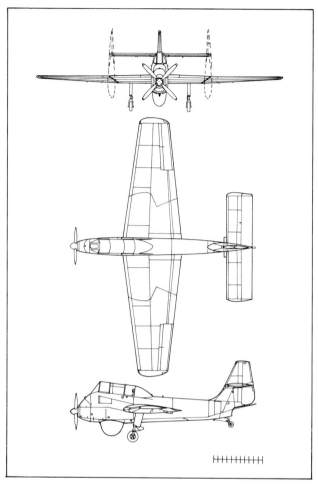

28   Seamew AS Mk 1

CHAPTER TWENTY-NINE
Supermarine Scimitar

Towards the end of World War Two consideration was being given to operating naval aeroplanes without undercarriages from carrier decks. The basis of the idea was that aircraft would be launched using a catapult and would land on a flexible rubber deck. It was envisaged that considerable savings in drag, weight and cost could be achieved by the exclusion of an undercarriage. As a consequence during this period several projects were offered without undercarriages and these included the Supermarine Type S.505 naval fighter project. The design was quite revolutionary with twin Rolls Royce AJ65 axial flow engines mounted in each side of the fuselage, thin straight wings and a butterfly form tailplane.

However before landing trials with a DH Vampire on a flexible deck could be carried out, the Admiralty had decided that there was a high level of uncertainty about this concept. Consequently it was considered prudent to revise the design of the S.505 project to enable it to operate from conventional carriers. To cover this

change of requirement Specification N9/47 was issued in September 1947 and Supermarine revised the design of the Type S.505, identifying it as the Type S.508: the essential change being the introduction of a retractable tricycle undercarriage. To accommodate this the wing thickness had to be increased, and although this did have an adverse effect on the aircraft's high speed performance it enabled a higher lift coefficient to be achieved which particularly improved its landing performance. The overall increase in weight required the wing area to be increased by 15 per cent and the unusual butterfly tail was retained by the Supermarine design team who considered it an ideal arrangement as it kept the tailplane clear of the jet effluxes.

Shortly after the issuing of Specification N9/47, a contract was placed for the manufacture of three Type S.508 prototypes. The first, VX133, flew for the first time at the A&AEE, Boscombe Down, on 31 August 1951 with M. J. Lithgow at the controls and subsequently deck landing trials were carried out on HMS *Eagle* during May and June of 1952. The second prototype, VX136, identified as the S.529 although virtually identical to the Type S.508, was involved in

Supermarine Type S 529, VX136, on board HMS *Eagle* for carrier trials in November 1953. (Vickers)

Line-up of Scimitar F Mk 1s of No. 700 X Squadron, Ford.

carrier trials on board *Eagle* in November 1953.

Following the success of Supermarine's research into swept wing aircraft a decision was taken to develop a swept wing version of the Type S.508 and the third S.508 was completed as the prototype of the swept wing version, being then identified as the Type S.525 and bearing the serial number VX138. Only the S.508 fuselage was retained for the Type S.525 which, as well as the new swept wings, was also fitted with a conventional vertical fin and rudder and mid-mounted tailplane replacing the butterfly tail. To achieve an acceptable low landing speed with the swept-wing version it was intended to introduce flap blowing, known by Supermarine as "super-circulation". This would maintain the boundary layer and increase the wing lift at low speed thereby reducing the landing speed. However, when XV138 flew for the first time in April 1954, it was not fitted with flap blowing, the plan being to introduce it at a later stage in the programme. In this configuration the S.525 was demonstrated at the SBAC Display at Farnborough in September of that year. VX138 was finally modified to introduce flap blowing during May 1955, but unfortunately was destroyed in a flying accident that July before the system could be fully assessed.

Scimitar F Mk 1, XD243/190/R, of No. 807 Squadron, HMS *Ark Royal*.

In early 1951 Supermarine received a contract for two prototypes, later increased to three, to meet Specification N113D. The company's design for this project was designated the Type S.544 although rather strangely the prototypes became known by the specification number N113, rather than by the company type number.

The first prototype N113, WT854, was flown by M. J. Lithgow at the A&AEE, Boscombe Down on 19 January 1956. Following the initial company trials WT854 was delivered to the RAE at Bedford in April 1956 for trials on the dummy deck which had been installed on the runway, and later proceeded to HMS *Ark Royal* for initial carrier trials. The two remaining N113 prototypes had both flown before the end of 1956 and the third prototype, WW134, the first to be fully equipped with flap blowing from the outset, carried out proving trials on board *Ark Royal* in January 1957.

During the development period the role of the N113 was changed from fighter to low-level strike aircraft with a capability to use the LABS attack system. This was an acronym for Low Altitude Bombing Sortie, a system of lobbing the bomb on to the target without having to approach too closely. The aircraft retained its standard four 30mm Aden cannons in the nose and was capable of carrying up to four 1000lb bombs under the wings. One of the Scimitar's claims to fame was that it was the first Fleet Air Arm aircraft with a nuclear strike capability.

Two Scimitar F Mk 1s, XD322/020/R and XD224/033/R, of No. 803 Squadron, ready for launching on the bow catapults of HMS *Ark Royal*. (B. J. Lowe)

Early in the development programme a production order for 100 aircraft to Specification N.113P1 was placed with Supermarine. However in the event only 76 production aircraft were completed. The final batch were presumably cancelled because the Buccaneer was by then entering service and taking over the strike role. The first production aircraft, XD212, flew for the first time on 11 January 1957, by which time the type had been named the Scimitar F Mk 1. The first Scimitars to enter Fleet Air Arm service were delivered to No. 700X Squadron, the Scimitar Intensive Flying Trials Unit (IFTU) which had formed at RNAS Ford on 27 August 1957 under the command of Cdr. T. G. Innes, AFC. On completion of the trials the IFTU disbanded on 29 May 1958 and was absorbed by No. 700 Squadron, with Cdr. Innes taking over command. No. 700 continued to operate Scimitars along with a variety of other types until February 1959.

The Scimitar entered first line service on 3 June 1958 when No. 803 Squadron commanded by Cdr. J. D.

Russell reformed at RNAS Lossiemouth with eight Scimitar F Mk 1s. After a period of training at Lossiemouth the squadron embarked in HMS *Victorious* on 25 September 1958. It immediately suffered a tragic loss when a Scimitar flown by the squadron commander broke its arrester hook on landing aboard. The aircraft ran over the side into the sea and despite desperate efforts to release him, the aircraft sank with its pilot still trapped in the cockpit. Lt. Cdr. G. R. Higgs took over command of the squadron and after a short spell of operations in Home waters, *Victorious* sailed for the Mediterranean where the squadron, in addition to operating from the carrier, spent a month operating from RNAS Hal Far on Malta before returning to the UK in January 1959.

No. 803 Squadron continued as part of *Victorious*'s carrier air group until February 1962, mainly in Home waters but with a Mediterranean cruise in October/November 1960 and serving in the Far East for most of 1961. In May 1962 No. 803 transferred to HMS *Hermes* operating both in the Mediterranean and the Far East during 1962 and 1963. After its final spell in *Hermes* the squadron returned to Lossiemouth in February 1964 and there its strength was increased to 16 Scimitars, when it took over aircraft from No. 800 which had then disbanded. In December 1964 the squadron joined

Scimitar F Mk 1 (Tanker) XD274/114/E of No. 800B Squadron, on the waist catapult of HMS *Eagle*, in 1966. (A. E. Hughes)

HMS *Ark Royal* and after two spells on board operating in Home waters, *Ark Royal* embarked her squadrons on 17 June 1965 and sailed for the Far East where she spent the next 12 months, her duties including two periods on the Beira patrol. After its return to the UK, the squadron disbanded at Lossiemouth on 1 October 1966 having operated the Scimitar continuously for over eight years.

The second Scimitar squadron was No. 807, commanded by Lt. Cdr. K. A. Leppard which had reformed with eight Scimitar F Mk 1s at RNAS Lossiemouth on 1 October 1958. The squadron remained land based until January 1960 when it spent a few days on board HMS *Victorious* for deck landing practice. Then, in March 1960, the squadron embarked in HMS *Ark Royal* for a Mediterranean cruise lasting some six months and in April of the following year flew aboard HMS *Centaur* for operations in the Mediterranean and Persian Gulf. Returning from the Mediterranean in May 1962 *Centaur* disembarked 807 to Lossiemouth on 14 May and the squadron disbanded the following day.

No. 800 Squadron commanded by Lt. Cdr. D. P. Norman AFC was reformed at Lossiemouth on 1 July 1959 with eight Scimitar F Mk 1s to form part of *Ark Royal*'s air group. Embarking in *Ark Royal* on 3 March 1960, the squadron was to spend much of the next four years operating from this ship in the Mediterranean and Far East. After finally returning to Lossiemouth in December 1963 the squadron disbanded on 25 February 1964, transferring its aircraft to No. 803.

This was, however, not the end of 800 Squadron's involvement with the Scimitar, for having reformed with Buccaneer S Mk 1s in March 1964 a separate

flight, No. 800B, was formed on 9 September 1964 equipped with four Scimitar F Mk 1s, which were to act as 'buddy' tankers for the squadron's Buccaneers. It was found that the Buccaneers were unable to take-off at a maximum weight so consequently they were launched with a limited amount of fuel on board meeting the Scimitar tankers shortly after take-off to top up their fuel. No. 800B Flight operated as an integral part of 800 Squadron for the next two years spending much of this time aboard HMS *Eagle*. The flight was finally disbanded towards the end of 1966 when the squadron re-equipped with the more powerful Buccaneer S Mk 2s, making the requirement for tanker aircraft unnecessary.

The last of the first line squadrons to receive Scimitar F Mk 1s was No. 804 which, commanded by Lt. Cdr. T. V. G. Binney, reformed at Lossie-mouth on 1 March 1960 with six aircraft. This squadron was however only short lived, being disbanded at Lossiemouth on 15 December 1961. During its short existence 804 spent some 11 months at sea in HMS *Hermes* including four months in the Far East.

To provide support for the operational squadrons, No. 736 Squadron, commanded by Lt. Cdr. J. D. Baker, re-equipped with Scimitar F Mk 1s in June 1959 at RNAS Lossiemouth to under-take the task of training pilots to operational first line standard. The course covered photo-reconnaissance, ground attack – including the use of air-to-ground Bullpup missile – and interceptions using guns and the air-to-air Sidewinder missile. As the Scimitars started to be phased out of first line service, No. 736 Squadron was disbanded at Lossiemouth on 26 March 1965.

This however was not the end of the Scimitar for as aircraft became surplus they were transferred to Airwork's civilian operated Fleet Requirements Unit (FRU) at Hurn. Training was needed for the Airwork pilots and to facilitate this, No. 764B Squadron was formed on 26 March 1965 from the remnants of 736 which had disbanded on the same day. On completion of the training task 764B was disbanded on 23 November 1965. The Scimitars entered service with the FRU in December 1965 and continued to be operated by the Unit until they were replaced by Hunters and Canberras in December 1970.

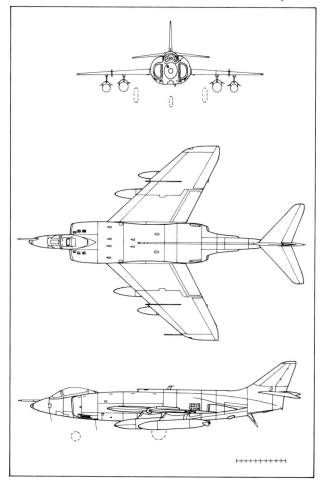

29 Scimitar F Mk 1

PRODUCTION

S.508 VX133.
S.529 VX136.
S.525 VX138.
S.544 WT854, WT859, WW134.
F. Mk 1 XD212 – XD250, XD264 – XD282, XD316 – XD333.

CHAPTER THIRTY
Hawker Hunter

At the time the Admiralty authorised the development of the Hawker P1040, as a ship-borne intercepter fighter for the Fleet Air Arm, the Air Ministry decided that this aircraft was unlikely to show any significant improvement over the RAF's current fighter the Gloster Meteor F Mk 4, and consequently withdrew from the project. However, in parallel with the development of the P1040 into the Fleet Air Arm's superb Sea Hawk, Hawker's chief designer, Sydney Camm, was working towards improved performance by the introduction of swept wings and more powerful engines.

The first step was the P1052, which was basically a swept wing version of the P1040. Two P1052 development aircraft, VX272 and VX279, were built, and although they did suffer from stability problems, they showed a marked improvement over the P1040. The design department by this time had come to the conclusion that the design should be based on the new Rolls Royce AJ65 engine, later to be named Avon, which would require the replacement of the bifurcated jet pipe by the more orthodox straightthrough jet pipe. At the same time they introduced further aerodynamic refinements, including swept-back tail surfaces. Discussions between the Ministry of Supply and the Hawker Aircraft Company led to the issue of Specification F.3/48. To meet this specification Hawker produced several design studies, one of which, P1067, was selected for development, and in 1948 Hawker received a contract for three prototypes. In the mean-

time, to speed development, the second P1052, VX279, was modified to introduce a straight through jet pipe and the swept tail surfaces. Unfortunately VX279 crashed on 3 April 1951 killing Squadron Leader T. S. (Wimpey) Wade, Hawker's chief test pilot. Shortly after this incident, on 20 July 1951, the first prototype P1067, WB188, made its maiden flight with the new chief test pilot, Squadron Leader Neville Duke, at the controls. By the time of this first flight Hawker had received a contract to build 198 P1067s for the RAF and, shortly afterwards, the type was officially named the Hunter.

In 1953 design work started on a two-seat trainer version of the Hunter for advanced training by the RAF. This resulted in a new nose being introduced which could accommodate the two pilots in side-by-side seating. Specification T 157D was issued early in 1954 and shortly afterwards a contract was placed for two prototypes and the first of these, XJ615, flew on 8 July 1955. A contract was subsequently placed for 55 of this new two-seat version identified as the Hunter T Mk 7, which was basically a F Mk 4 fitted with the two-seat trainer nose. The first production Hunter T Mk 7, XL563, was completed late in 1957 and the last of the batch had been delivered by February 1959.

In 1957 it was apparent that the Sea Fury T Mk 20 and the Sea Vampire T Mk 22 advanced trainers would not be suitable for training the next generation of Fleet Air Arm pilots for the Sea Vixens and Scimitars that were expected soon to be entering service. As the only transonic trainer in production for the RAF at the time, it was virtually inevitable that the Hunter T Mk 7 would be selected by the Fleet Air Arm for this role.

The first Hunter T Mk 8 built for the Fleet Air Arm, XL580, in the special blue and white colour scheme when used by the Flag Officer Flying Training (FOFT) in 1962. (B.Ae.)

Hunter T Mk 8, XL598/870, of FRADU in 1982. The camouflaged fin with a fin flash suggests that it has been fitted with a replacement ex-RAF fin. (B.Ae.)

Initially, to carry out the development to meet the Royal Navy requirement, a Hunter F Mk 4, WW664, was converted by Hawker's into the prototype of the Royal Navy's Hunter T Mk 8 which differed from the RAF's T Mk 7 primarily by the introduction of an airfield arrester hook and a brake parachute. The Hunter T Mk 8 was cleared for Fleet Air Arm service following trials, using WW664, at the A&AEE, Boscombe Down early in 1958. Initially, ten of the RAF's production order for 55 Hunter T Mk 7s were converted on the production line to T Mk 8 standard for the Royal Navy, and subsequently a contract was placed for the conversion of 18 Hunter F Mk 4s into Hunter T Mk 8s, the work being carried out by Hawker Aircraft Ltd, and Sir W. G. Armstrong Whitworth Ltd.

Deliveries of Hunter T Mk 8s to the Fleet Air Arm started on 1 July 1958 with the early aircraft being delivered to RNAS Lossiemouth, where they joined No. 736 Squadron, the Naval Air Fighter and Strike School, commanded by Lt. Cdr. L. E. A. Chester-Lawrence. The first Hunter received by No. 736 Squadron, XL581, was destroyed in a crash at Lossiemouth not long after it had been delivered. Shortly afterwards No. 736 handed over its Hunters to No. 764 Squadron, commanded by Lt. Cdr. D. T. McKeown. This squadron, which was responsible for Air Warfare Instructor Training and Swept Wing Conversion, received its full

Hunter T Mk 8M, XL580/719/VL, with a Sea Harrier FRS Mk 1 and a Harrier T Mk 4N of No. 899 Squadron, Yeovilton. (B.Ae.)

complement of 12 Hunter T Mk 8s by August 1959, and was to remain the prime Fleet Air Arm operator of the type for many years until disbanded in July 1972.

A further 15 Hunter F Mk 4s were converted into Hunter T Mk 8s in 1963/1964. The first eleven, identified as Hunter T Mk 8Cs, were fitted with an interim version of the Tactical Air Navigation System, TACAN. The remaining four, identified as Hunter T Mk 8Bs, were fitted with the full TACAN. Subsequently, in 1978, two Hunter T Mk 8s, XL602 and XL603, were modified to carry Ferranti Blue Fox radar as part of the Sea Harrier FRS Mk 1 development programme, and were identified as the Hunter T Mk 8M. A third aircraft, XL580, was later brought up to T Mk 8M standard. On completion of the development programme, the aircraft were transferred to RNAS Yeovilton, where they are used for systems training by No. 899 Squadron, the Sea Harrier Headquarters Squadron.

During their relatively short existence of an average eleven months, Nos. 700B, 700Y and 700Z Squadrons, which made up the Intensive Flying Trials Units, used their Hunter T Mk 8 to train aircrew for the Buccaneer S Mk 2, Sea Vixen FAW Mk 1 and Buccaneer S Mk 1 respectively. On 1 August 1963 No. 759 Squadron commanded by Lt. Cdr. C. D. W. Pugh, MBE, reformed at RNAS Brawdy with Hunter T Mk 8 and T Mk 8C aircraft. No. 759 was the Fleet Air Arm's Advanced Flying Training School, converting Jet Provost trained pilots to the Hunter. For its excellent work, No. 759 was awarded the annual Boyd Trophy in 1965. The squadron disbanded on 24 December 1969.

Following the success of the Hunter T Mk 8 in Fleet Air Arm service, it was decided that the single-seat

Hunter GA MK 11, XF977/691/LM, of No. 764 Squadron, Lossie-mouth. (via R. C. Sturtivant)

Hunter would make an ideal weapons trainer. Consequently, in 1961, a contract was awarded to Hawker Aircraft for the supply of 40 Hunter GA Mk 11s. This was basically a variant of the RAF's F Mk 4 modified by the removal of the Aden guns and guns pack and the introduction of the TACAN navigational system and an airfield arrester hook under the rear fuselage. The first conversion, XE712, was used for type clearance trials at A&AEE Boscombe Down during April/May 1962 before being delivered to the Fleet Air Arm. The Hunter GA Mk 11 entered squadron service in June 1962 when No. 738, an Advanced Training Squadron, commanded by Lt. Cdr. F. Hefford, DSC, re-equipped at RNAS Lossiemouth with both versions of the Royal Navy's Hunters, the GA Mk 11s and the T Mk 8s. The squadron moved to RNAS Brawdy on 1 January 1964, where it remained until it was disbanded on 8 May 1970. The only other squadron to operate the single-seat Hunters was No. 764, which received its first GA Mk 11s during July 1962 to operate alongside its existing equipment, the Hunter T Mk 8. No. 764 also

subsequently operated a small number of Hunter PR Mk 11s, a photographic reconnaisance version of the Hunter GA Mk 11, which had been converted by Short Brothers at Belfast. When the squadron disbanded at Lossiemouth on 27 July 1972 it had a complement of ten Hunter GA Mk 11s and four Hunter T Mk 8Cs.

Early in 1969 the civilian operated Fleet Requirements Unit at Hurn, run by Airwork Ltd, received a small number of Hunter T Mk 8s and GA Mk 11s. The following year the Air Direction Training Unit, run at Yeovilton by Airwork Ltd, also took delivery of a quantity of Hunter T Mk 8s and GA Mk 11s. These two units combined on 1 December 1972 at Yeovilton, becoming the Fleet Requirements and Air Direction Training Unit (FRADTU), although later the 'Training' was deleted from their title. FRADU initially operated Canberras, Sea Vixens and Hunters. Although the Sea Vixens were withdrawn from service in 1974 and the contract to operate the unit was taken over by Flight Refuelling Aviation Ltd in 1983, the unit continues to operate Canberras and Falcon 20s along with Hunter T Mk 8Cs and Hunter GA Mk 11/PR Mk 11s. Falcon 20s have taken over some of the Hunter's roles, however a number of Hunters are expected to remain in service for the foreseeable future.

Two Hunter GA Mk 11s, WV267/836 and XE689/864 of FRADU.

Two Hunter GA Mk 11s, WT744/868/VL and XF977/865/VL, of
FRADU, Yeovilton, in the new all grey colour scheme.
(Flight Refuelling)

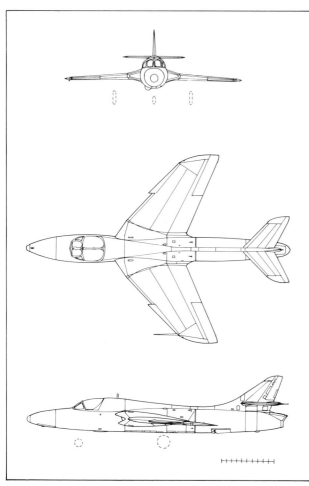

30   Hunter T Mk 8

PRODUCTION

T Mk 8   XL580 – XL582, XL584, XL585, XL598,
XL599, XL602 – XL604, WT701, WT702, WT722,
WT745, WT755, WT772, WT799, WV319, WV322,
WV363, WW661, WW664, XE664, XE665, XF289,
XF322, XF357, XF358.

T Mk 8B   XF967; XF978, XF995.

T Mk 8C   WV396, WV397, XF938, XF939, XF942,
XF983, XF985, XF991, XF992, XF994, XL604.

GA Mk 11   WT711 – WT713, WT718, WT721,
WT723, WT741, WT744, WT804 – WT806, WT808 –
WT810, WV256, WV257, WV267, WV374, WV380 –
WV382, WW654, WW659, XE668, XE673, XE674,
XE680, XE682, XE685, XE689, XE707, XE712,
XE716, XE717, XF291, XF297, XF300, XF301,
XF368, XF977.

CHAPTER THIRTY-ONE
De Havilland Sea Vixen

De Havilland put forward proposals to the Admiralty in 1946 for an advanced all-weather fighter which had been allocated the de Havilland project number DH110. The project continued with the, then, customary de Havilland twin boom layout, but in this case the aircraft was to be powered by two engines and have swept wings. The pilot's cockpit was offset to port, with the observer being 'buried' in the fuselage beside the pilot; access being provided by way of a flush fitting hatch.

Specifications Nos. N.40/46 and F.44/46 were issued in January 1947 to cover requirements for night fighters for the Fleet Air Arm and RAF respectively. It was considered that as the basic requirement of both specifications was very similar, the same basic airframe could be used; modified as necessary to meet the Fleet Air Arm requirement for carrier operation. The RAF requirement was updated early in 1948, a new Specification, F.4/48 being issued and the following year the Fleet Air Arm specification was revised and re-issued as N.14/49. Of the designs submitted, de Havilland's DH110 was selected as the prime project with the delta

The partially navalised third prototype DH110, XF828, during carrier trials aboard HMS *Ark Royal* in April 1956. (B.Ae.)

wing Gloster GA.5 as a back up. Initially an order was placed in April 1949 for seven night fighter and two long range day fighter prototypes for the RAF and two night fighter and two strike fighter prototypes for the Royal Navy. At the same time four prototypes of the Gloster GA.5 to the RAF F.4/48 specification were ordered. In 1949, for mainly financial reasons – although there were some doubts in certain Royal Navy circles about operating swept winged aircraft from carrier decks – the Royal Navy cancelled the DH110 prototypes having decided to order the Sea Venom as a replacement for the Sea Hornet. At the same time the orders for the land based DH110 and the GA.5 were reduced to two off each.

Work on the two DH110 prototypes continued at the de Havilland factory at Hatfield and the first aircraft, WG236, flown by John Cunningham took to the air on 26 September 1951, followed by the all-black second prototype, WG240, on 25 July 1952.

Tragedy struck on 6 September 1952, at the SBAC Display at Farnborough, when WG236 disintegrated as it flew low and fast towards the spectators, killing the pilot John Derry and his observer Tony Richards along with 29 spectators who were struck by one of the engines as it landed in the crowd. The second proto-

Late in the 1950s, the Royal Navy had changed its aircraft mark numbering system so that Fleet Air Arm aircraft were no longer numbered from Mk 20 onwards but were numbered within a common system with the RAF. As a result of this the Sea Vixen FAW Mk 20 was re-designated as the FAW Mk 1. Design of an improved version of the Sea Vixen started in the early 1960s and this exercise resulted in the introduction of pinion tanks forward of the tail booms. This served the dual purpose of increasing the internal fuel tankage and providing accommodation for the equipment necessary for the new Red Top air-to-air missile systems. Fairings were fitted running along the tail booms from the rear of the new pinion tanks. This new version was designated the Sea Vixen FAW Mk 2 and two FAW Mk 1s, XN684 and XN685, were diverted from the production line at Christchurch to Hatfield for conversion into the FAW Mk 2 prototypes. XN684 flew for the first time as a Mk 2 on 1 June 1962 in the hands of Chris Capper, Christchurch's chief test pilot.

Following a decision to close the factory at Christchurch, Sea Vixen production was transferred to de Havilland's factory at Chester, where a single Mk 1, XP918, was built, followed by all 29 of the production Mk 2s. The first production Mk 2, XP919, flew for the first time at Chester on 8 March 1963 and Sea Vixen production came to an end with the first flight of the last production aircraft, XS590, on 3 February 1966. In addition to the production Mk 2s, a number of FAW Mk 1s were converted to Mk 2 standard, some at Chester and others at the Royal Naval Air Yard at Sydenham in Northern Ireland. The two prototypes and the first two production Mk 2s were used for development trials, the former being involved in Red Top trials initially at Hatfield and later at the A&AEE Boscombe Down. Both prototypes were subsequently brought up to service standard at Chester before being delivered to the Fleet Air Arm in 1966. XP919 was used for a variety of trials at Hatfield and Boscombe Down, including trials with the Bullpup air-to-ground missile, while XP920 was used for service trials, including a period early in 1964 when it operated from HMS

Hermes. In 1969 it was used for development of the frangible hatch over the observer's position, which allowed the observer to eject directly through the hatch without first having to jettison it.

Deliveries of the Sea Vixen FAW Mk 2s to the Fleet Air Arm started with the third production aircraft, XP921, delivered to the Aircraft Handling Unit at RNAS Brawdy on 13 August 1963. The Mk 2 started to enter first line service in February 1964, when early production aircraft were received by the Sea Vixen Headquarters Squadron, No. 899 at Yeovilton. On 15 June 1964 the squadron, commanded by Lt. Cdr. D. C. Matthews, was re-commissioned as the Sea Vixen FAW Mk 2 Intensive Flying Trials Unit, although it continued to operate a small number of Mk 1s until the following September. The squadron embarked in HMS *Eagle* on 2 December 1964, and sailed for the Far East, where it was to spend the next 18 months, except for a short period in the summer of 1965 when the squadron flew back to Yeovilton for some home leave. During the latter stages of its Far East cruise, *Eagle* was diverted from exercises off Malaysia to take a leading part in the Beira Patrol, an attempt by the British Government to prevent oil supplies from reaching Rhodesia after its illegal Unilateral Declaration of Independence. Until it disbanded at Yeovilton on 23 January 1972, No. 899 Squadron was a permanent element of *Eagle*'s air group and spent periods operating in both Home and Far Eastern waters.

The All Weather Fighter Training Squadron, No. 766 at Yeovilton, received its first Sea Vixen FAW Mk 2s in July 1965, but continued to operate both Mk 1s and Mk 2s until May 1968, when the last of the Mk 1s was phased out. The squadron disbanded on 10 December 1970, when with plans formalised for the run down of operational Sea Vixen squadrons, there was no longer a requirement to train new Sea Vixen aircrew.

No. 893 Squadron, commanded by Lt. Cdr. G. P. Carne, reformed on 4 November 1965, with eleven Sea Vixen FAW Mk 2s, and shortly afterwards, in December 1965, No. 892 Squadron commanded by Lt. Cdr. J. N. S. Anderson, reformed with a small number of Sea Vixen FAW Mk 2s, the re-equipment continuing progressively until the full complement of twelve aircraft was reached. No. 892 was fairly short lived as a Sea

Sea Vixen FAW Mk 2, XN697/137/E, of No. 899 Squadron on HMS *Eagle* in April 1970.

Sea Vixen FAW Mk 2, XP957/137/E, of No. 899 Squadron on the waist catapult of HMS *Eagle* in 1966. (A. E. Hughes)

Vixen FAW MK 2 squadron, disbanding on 4 October 1968 after serving as an element of the air group in HMS *Hermes*, until replaced by No. 893 in May 1968.

During its last year, No. 892 formed an aerobatic team of six aircraft known as 'Simon's Circus', named after the squadron's C.O. Lt. Cdr. Simon Idiens, and this team gave an excellent display at the 1968 SBAC Show at Farnborough.

No. 893 Squadron had been part of the air group in HMS *Victorious* until the decision was taken in November 1967 to scrap the ship following a minor fire on board while it was berthed at Portsmouth. The squadron then transferred to HMS *Eagle* as part of her air group for exercises in the Mediterranean and the Far East, before being disbanded on 14 July 1970. The final squadron to re-equip with Sea Vixen FAW Mk 2s was

No. 890 commanded by Lt. Cdr. M. F. Kennett, which reformed on 14 August 1967 with six aircraft. This squadron operated as the Sea Vixen FAW Mk 2 Headquarters Squadron, carrying out operational trials and providing some operational training, in addition to providing aircraft and crews for fleet requirement and air-to-air tanking duties. These latter tasks were subsequently taken over by the Fleet Requirement Unit run by Airwork at Yeovilton. No. 890 Squadron finally disbanded at Yeovilton on 6 August 1971. The final operator of the Sea Vixen was the civilian Air Direction Training Unit run at Yeovilton by Airwork Ltd and operating a number of FAW Mk 2s from January 1971 until November 1972.

Early in the 1960s air-to-air refuelling trials were carried out using XJ488 and XJ516. The system of refuelling developed involved the carrying of a Flight Refuelling Mk 20 pod under the starboard wing and a probe on the port wing. Known as the "buddy" refuelling system, it proved to be an effective means of extending the range of operational aircraft or alternatively permitting them to take off with the maximum load of weapons. The aircraft took off with only a limited amount of fuel to keep its all-up weight under the maximum permissible for take-off, the fuel then being topped up by air-to-air refuelling. Once developed, this system was used extensively by the Fleet Air Arm's Sea Vixens and also by Scimitars and Buccaneers.

Development of a target tug version was carried out initially using the Sea Vixen FAW Mk 1, XJ560, which towed the Del Mar target. On completion of the trials XJ560 was converted to Sea Vixen FAW Mk 2 standard and return to service.

In the early 1970s the Sea Vixen was selected for conversion into an unmanned target drone to be known as the D Mk 3. As the Sea Vixen FAW Mk 2s were being phased out of service, five were delivered to RAE Llanbedr for conversion, while many of the remainder were delivered to the RAE at Farnborough for storage and preparation for the conversion programme, which was to be carried out by Flight Refuelling Ltd at Tarrant Rushton. However, after only a few aircraft had been converted to Sea Vixen D Mk 3 standard, the programme was abandoned because of a shortage of funds.

Flight Refuelling Ltd used Sea Vixens for trial purposes and it was Sea Vixen FAW Mk 2 XJ580 which was used for the development of the Mk 32 refuelling pod for the VC10 tankers.

The Sea Vixen had given excellent service to the Fleet Air Arm, but had been retired prematurely following political decisions to bring the Fleet Air Arm's fixed wing flying to an end. At the time of their withdrawal, the airframes still had at least ten years fatigue life left and there was still potential for further systems development.

PRODUCTION

Prototypes   WG236, WG240, XF828.
FAW Mk 1   XJ474 – XJ494, XJ513 – XJ528, XJ556 – XJ586, XJ602 – XJ611, XN647 – XN658, XN683 – XN710, XP918.
FAW Mk 2   XP919 – XP925, XP953 – XP959, XS576 – XS590.

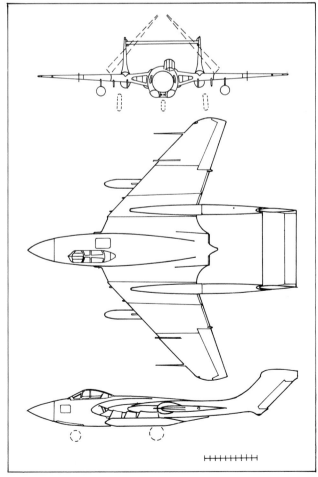

31   Sea Vixen FAW Mk 1

CHAPTER THIRTY-TWO
English Electric Canberra

The Canberra, a twin jet-engined long range bomber, was designed by the English Electric Company to meet Specification B3/45 and was the first jet powered bomber to enter service with the RAF. The first operational version of the Canberra was the B Mk 2 which entered service with No. 101 Squadron in May 1951. The Canberra B Mk 2 was built in large quantities for the RAF: 205 built by English Electric and a further 225 by sub-contractors.

The Fleet Air Arm's initial involvement with the Canberra was fairly brief and on a very small scale, starting in May 1961 with the delivery of a small number of Canberra U Mk 14s to supplement the Firefly U Mk 9s, Meteor U Mk 15s and U Mk 16s operated by No. 728B Squadron at Hal Far in Malta. The Canberra U Mk 14 (later D Mk 14), the Royal Navy equivalent of the Royal Air Force's U Mk 10, was a conversion of the RAF's Canberra B Mk 2 which was

modified to operate in the unmanned role being radio controlled from the ground. A total of six Canberra U Mk 14s were converted by Short Brothers at Belfast, and this involved the introduction of radio control equipment and a servo-hydraulic control system similar to that used for the Canberra PR Mk 9.

No. 728B Squadron, commanded by Lt. Cdr. J. G. Corbett, was the Fleet Air Arm's Pilotless Drone Target Unit, and at the time was heavily involved in providing targets for the Seaslug ship-to-air missile trials that were being carried out from the missile trials ship HMS *Girdle Ness* and it was in this role that the Canberras were used. The squadron was disbanded on 2 December 1961 following the successful completion of the trials.

It was several years before the Canberra was to make a further appearance in Fleet Air Arm service, this time in the target-towing role as a replacement for the obsolescent Meteor TT Mk 20. The TT Mk 20 had a

Canberra TT Mk 18, WJ614/846, of FRADU, Yeovilton. (HMS *Heron*)

maximum operational ceiling of 10,000 ft, an endurance of under two hours and a limited equipment carrying capability. It was obvious that these limitations meant a replacement had to be found. Early in 1967 the Canberra was selected as the replacement and the first aircraft, a B Mk 2, was delivered to the British Aircraft Corporation (BAC) factory at Samlesbury for conversion into the first Canberra TT Mk 18. Following trials by BAC and the A&AEE at Boscombe Down, a batch of ten Canberra B Mk 2s was converted into TT Mk 18s for the Royal Navy. The first aircraft, WK123, was delivered to the Fleet Requirements Unit (FRU), run by Airwork at Hurn, on 15 September 1969. The Canberras served alongside the unit's Meteor TT Mk 20s until March 1971 when the last Meteor was retired from service. In addition to the TT Mk 18s the unit also operated two Canberra T Mk 4s for crew training. On 1 December 1972 the FRU transferred to RNAS Yeovilton where it combined with the Air Direction Training Unit (ADTU) to become the Fleet Requirements and Air Direction Training Unit (FRADTU). In addition to the Canberras the unit operated Hunters and Sea Vixens. The Sea Vixens were phased out of service in 1974 – about the same time that the word "Training" was dropped from the unit's title, which became FRADU.

The Canberra TT Mk 18 soon proved to be a significant improvement over the Meteor TT Mk 20 with an operational ceiling of 15,000ft and an endurance of about three hours. Its primary duty was to tow sleeve type targets for gunnery practice for Royal Navy warships. For optical or radar tracking the Rushton type targets were used. Since the introduction of the

Two Canberra T Mk 22s, WH801/850 and WH780/853, of FRADU, Yeovilton. (HMS *Heron*)

Sea Harrier in 1979 the Canberras have also been used to tow banner type targets for air-to-air firing practice.

The final version of the Royal Navy's Canberra was a training variant, the T Mk 22, which started to enter service with FRADU early in 1974 as a replacement for the Sea Vixen FAW Mk 2s. The T Mk 22 was a conversion of the RAF's Canberra PR Mk 7 carried out by BAC at their Samlesbury factory, the major change being the introduction of the long pointed nose to accommodate the "Blue Parrot" radar. A total of seven Canberra T Mk 22s were converted at Samlesbury and all these were delivered to FRADU where they are used to simulate anti-ship missile attacks on Royal Navy ships, enabling their crews to train in the defensive action necessary to counter such an attack. The T Mk 22s are also used to undertake various radio and radar calibration duties.

In 1983 the contract for the management and staffing of FRADU came up for renewal and was offered on competitive tender. The competition was won by Flight Refuelling Aviation Ltd, who took over the operation of FRADU from Airwork.

At the time of writing civil registered Dassault Falcon 20s have replaced the Canberra T Mk 22s and are starting to replace the Canberra TT Mk 18s of FRADU and it is expected that the last of the Fleet Air Arm's Canberras will be replaced in the near future.

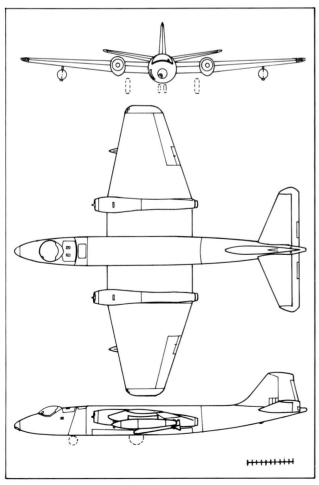

32   Canberra TT Mk 18

PRODUCTION

T Mk 4   WJ854, WJ866.

U Mk 14/D Mk 14   WD941, WH704, WH720, WH876, WH921, WJ638.

TT Mk 18   WE122, WH856, WH887, WJ574, WJ614, WJ636, WJ717, WK123, WK126, WK142.

T Mk 22   WH780, WH797, WH801, WT803, WT510, WT525, WT535.

CHAPTER THIRTY-THREE
Westland/Sikorsky Wessex

Although both the Dragonfly and Whirlwind had in turn been major steps forward in operational helicopter design, neither had been particularly successful in the anti-submarine role. It was considered that a larger, more powerful helicopter, ideally powered by two engines, would be needed to satisfy the ASW requirement. In the early 1950s the Admiralty had two helicopters under consideration for this role, these being the American Bell HSL-1, a helicopter designed specifically for the ASW role, and the twin rotor Bristol 173 which had been designed as an experimental transport helicopter to meet Specification E.4/47. During 1953 an order was placed for 18 Bell HSL–1s but a shortage of foreign currency resulted in the order being cancelled the following year. After evaluation by

Wessex HAS Mk 1, XP104/B/A, of No. 845 Squadron, HMS *Albion*, off-loading Royal Marine Commandos during an exercise.

the RAF the Bristol 173 Mk 1 XF785 was used for carrier trials on HMS *Eagle* during 1953 and although these were completed successfully they did show that the Bristol 173 would require some significant modifications before it would be suitable for carrier operations. These included a reduction in overall length to improve deck handling and a longer undercarriage to allow for the carriage of a torpedo under the fuselage. Specification HAS.107D was issued to cover the Royal Navy requirement and to meet this Bristol proposed the Type 191 which was a redesign of their Type 173 project.

In April 1956 a contract was placed on Bristol for the manufacture of three prototypes and 65 production Type 191s for the Royal Navy. As an interim measure, to enable naval trials to get under way, a Type 173 Mk 5 was built which featured a shortened fuselage and a

Wessex HAS Mk 1, XS872/061/E, of No. 820 Squadron, HMS *Eagle.*
(RNAS Culdrose)

long stroke undercarriage. However by the end of 1956
it was realised that engine and transmission problems
then being encountered by the Type 173s were not
going to be easy to resolve and that this was likely to
put back the in-service date of the Type 191 to an
extent that would be unacceptable to the Royal Navy.
Consequently the contract for the Type 191 was
cancelled, leaving the Fleet Air Arm ASW units to
struggle on with their somewhat inadequate Whirlwinds.

At about the same time the US Navy had also
abandoned the Bell HSL-1 project, and as a replacement for its anti-submarine Whirlwinds, Sikorsky had
produced the S.58 which had flown for the first time in
1954. With the sudden urgent need to find another
ASW helicopter for the Fleet Air Arm following the
cancellation of the Bristol Type 191 the Royal Navy
approached Westland to provide a helicopter to fill the
gap. Realising the urgency of the situation, Westland
proposed a licence built version of the Sikorsky S.58 to
be powered if possible by British built gas turbine
engines. Agreement to build the S.58 under licence was
reached by the end of 1956 and although Westland had
intended to power it with twin gas turbine engines the
urgency of the situation resulted in Westland offering
the S.58 with a single 1250shp Napier Gazelle engine.

In the meantime Westland had purchased a standard
US Navy type S.58 from Sikorsky. This was delivered
to Yeovil in June 1956, re-assembled and flown for the
first time in the UK on 24 June, with the class "B"
markings G-17-1. Being powered by the 1525hp Wright
R-1820-84 piston engine, it was not fully representative
of the projected Westland version of the S.58. However
in July the aircraft received serial number XL722 and
was delivered to the A&AEE, Boscombe Down for
evaluation, and demonstrations to the British services,
moving in August to RNAS Lee-on-Solent for initial
service trials by the Royal Navy. By the end of 1956
work had been put in hand to install an early
production Napier Gazelle N.Ga.11 in the nose of
XL722 which flew for the first time in this configuration
on 17 May 1957. Initially problems were encountered
with the intakes and engine cooling but these were soon
resolved and in September the N.Ga.11 engine was
replaced by the 1450shp N.Ga.13, and it was in this
form that the aircraft made its first public appearance at

the 1957 SBAC Display at Farnborough.

Following the success of the trials with the turbine
engined XL722, an order was placed for two pre-
production WS-58s which by then had been named the
Wessex HAS Mk 1, and the first of these, XL727, flew
on 20 June 1958. The HAS Mk 1 was designed to be
operated by a crew of three comprising, pilot, observer
and sonar operator but for night operations a second
pilot was also carried. The Wessex saw the introduction
of a wide range of improved systems including an automatic flight control system, navigational radar and
radio altimeter. For operation of the dunking sonar it
was essential for the helicopter to hover at the required
height, a particularly difficult manoeuvre to achieve at
night, but made possible by the various control systems. The normal armament consisted of machine
guns, air-to-surface rocket projectiles and one or two
homing torpedoes carried on pylons mounted on each
side of the fuselage, although these could be replaced
by four Nord SS.11 wire guided missiles.

Contracts were placed with the Westland Aircraft
Company for 140 Wessex HAS Mk 1s for the Royal
Navy, including three prototypes and a development
batch of nine aircraft. Deliveries commenced early in
1960 and on 1 April No. 700H Flight reformed as the
Wessex Intensive Flying Trials Unit at RNAS Cul-

Wessex HAS Mk 3, XM918/585/CU, of No. 700H Squadron, the
Wessex Intensive Flying Trials Unit, Culdrose, landing on the
destroyer HMS *Hampshire* in 1967. (HMS *Daedalus*)

Wessex HU Mk 5, XT461/V-R/B, of No. 846 Squadron, HMS *Bulwark*. in 1979. (B.Ae.)

drose, under the command of Lt. Cdr. R. Turpin. The flight was disbanded in January 1962 following the satisfactory completion of the trials. The Wessex entered first line service on 4 July 1961 when No. 815 Squadron, under the command of Lt. Cdr. A. L. L. Skinner, reformed at RNAS Culdrose with eight HAS Mk 1s. The squadron embarked in HMS *Ark Royal* on 13 November for the Mediterranean, but engine problems with the Wessex resulted in the squadron disembarking at Malta and returning to the UK aboard HMS *Victorious*, for modifications to the engines. In March 1962 the squadron re-embarked in HMS *Ark Royal* which sailed to the Far East by way of the Mediterranean, returning to the UK in December 1962. For its work in bringing the Wessex HAS Mk 1 into service and pioneering night and all-weather ASW tactics and Wessex ASR procedures No. 815 Squadron was awarded the Boyd Trophy for 1962. The squadron returned to the Far East aboard *Ark Royal* in 1963 and disembarked to Khormaksar in Aden on 19 December 1963 to provide support for the British forces involved in the disturbances there. However, with a rebellion taking place in Tanganyika, the squadron embarked in HMS *Centaur* where the Wessex were converted into commando helicopters, basically by the removal of the anti-submarine equipment, and were then used to put No. 45 Royal Marine Commando ashore in Tanganyika.

The squadron returned to Aden in May 1964, re-embarking on HMS *Centaur* in June to sail to Singapore, before returning to the UK in December 1964. Following further spells aboard *Ark Royal* on exercises in the Far East and Home waters, the squadron disbanded at Culdrose on 7 October 1966.

When No. 719 Squadron, the Joint Anti-Submarine School Flight at RNAS Eglinton, re-equipped on 5 October 1961 with four Wessex HAS Mk 1s, it was immediately given first line status, became No. 819 Squadron and operated as the Wessex headquarters squadron providing aircraft as required for exercises with the Joint Anti-Submarine School at Londonderry. The squadron also provided aircraft needed to operate from NATO carriers engaged on exercises and also to be based on ships of the Royal Fleet Auxiliary. Its Mk 1s were replaced by HAS Mk 3s in April 1968.

The last Wessex squadron to appear in 1961 was No. 814, commanded by Lt. Cdr. L. B. J. Reynolds, which reformed at Culdrose on 28 November with eight HAS Mk 1s. After working up it embarked in HMS *Hermes* for five months in the Mediterranean. Following its return to the UK in October 1962, the squadron re-embarked in HMS *Hermes* in November for a tour of duty in the Far East. Early in 1964, while still in the Far East, No. 814's Wessex were transferred to HMS *Victorious*. They were then stripped of their anti-submarine equipment in readiness for use as assault helicopters on the commando carrier HMS *Albion,* in support of anti-terrorist activities in East Africa. In the event they were not required and were restored to their anti-submarine role and returned to *Victorious* aboard which they returned home in the summer of 1965. Before No. 814 Squadron replaced its Wessex HAS Mk 1s with the later HAS Mk 3s in August 1967, it had completed a further tour of duty in the Far East aboard HMS *Victorious*.

The significant improvement of payload and range achieved by the Wessex, over the Whirlwind inevitably resulted in its use as the basic equipment for the commando helicopter squadrons. Initially a small number of Wessex HAS Mk 1s were modified for the commando role by the removal of the sonar and much of the specialised flight control system enabling the aircraft to carry up to 16 troops or, in the casualty evacuation role (CASEVAC), eight stretcher cases. Twelve of these modified HAS Mk 1s were used as the equipment of No. 845 Squadron which reformed on 10

Wessex HU Mk 5, XT475/624/PO, of No. 772 Squadron, Portland. (HMS *Osprey*)

April 1962 at Culdrose, under the command of Lt. Cdr. A. A. Hensher. The squadron embarked in HMS *Albion* in September for the Far East and was soon operating in support of the forces acting against the Indonesian backed rebels in Brunei. No. 845 carried out operations in and around Brunei for more than two years, before returning to the UK in September 1965. For its services in the defence of Malaysia, the squadron was awarded the Boyd Trophy for 1964, making it the second Wessex HAS Mk 1 squadron to win the trophy.

In March 1964 No. 829 Squadron, commanded by Lt. Cdr. K. M. Mitchell, DFC, was reformed at Culdrose as the headquarters squadron for the Wasps operating from the Royal Navy's frigates and the Wessex operating from the county class destroyers. The squadron retained responsibility for the destroyers' Wessex helicopters until they were taken over by No. 737 at Portland in June 1970. There were only two other first line squadrons that operated Wessex HAS Mk 1s, both in the ASW role, No. 820 which reformed at Culdrose in September 1964 and operated at sea from HMS *Eagle,* and No. 826 which had also reformed at Culdrose and operated from HMS *Hermes.* Both these squadrons had replaced their Wessex HAS Mk 1s by May 1969.

Development of the Wessex had continued at Westland and the first major development was the introduction of coupled 1350shp Bristol Siddeley Gnome Mk 110/111 turbines. This installation was first carried out on an early production Wessex HAS Mk 1, XM299, making it, in effect, the prototype of the HC Mk 2 which was subsequently put into production for the RAF. It was a development of the Wessex Mk 2 into an assault helicopter for the Royal Marines and identified as the Wessex HU Mk 5 that was to be built for the Fleet Air Arm. The HU Mk 5 was designed for commando assault duties from carriers and, in addition to carrying the same weapons as the HAS Mk 1, was also able to use the Nord AS12 wire guided anti-ship missile.

Wessex HU Mk 5, XS484/821/CU, of No. 771 Squadron, Culdrose, operating in the SAR role along the rugged Cornwall coastline. (RNAS Culdrose)

Wessex HU Mk 5, XT770/816, converted to the VIP role and operated by No. 781 Squadron, Lee-on-Solent, and nicknamed the 'Green Parrot', on HMS *Eagle* in April 1970. (Rolls Royce)

The Wessex HU Mk 5 prototype, XS241, flew for the first time on 31 May 1963 followed by the first production aircraft, XS479, on 17 November 1963. No. 700V Flight, the Wessex HU Mk 5 Intensive Flying Trials Unit commanded by Lt. Cdr. C. J. Isacke, formed at Culdrose on 29 October 1963 with six aircraft. The trials were completed successfully by 7 May 1964 when No. 700V Flight disbanded and became the nucleus of No. 848 Squadron which reformed on the same day under the command of Lt. Cdr. G. A. Andrews and was equipped with 18 Wessex HU Mk 5s.

With the introduction of the HU Mk 5 into service, the Fleet Air Arm formed an advanced and operational flying training commando helicopter squadron, identified as No. 707 Squadron. This squadron, under the command of Lt. Cdr. D. J. Lickfold, MBE, reformed at Culdrose on 9 December 1964 and although primarily a training squadron, it carried out a variety of other duties including communications, development, and weapons trials. During July 1974 the squadron was given the honour of providing a Wessex conversion course for HRH the Prince of Wales, and for this purpose a special "Red Dragon" Flight of two aircraft was formed and, on completion of the task, disbanded on 12 December 1974. No. 707 Squadron was given first line status on 19 April 1982 becoming No. 848 and sailing with the Falklands Task Force. No. 707 Squadron reformed the following month with Wessex HU Mk 5s to continue its training task.

Starting with No. 848 Squadron, a total of five first line squadrons were to operate Wessex HU Mk 5s. Three of these, Nos. 848, 845 and 847, were involved in operations in Malaysia, operating from either the commando carriers HMS *Albion* and HMS *Bulwark,* or the assault ships HMS *Fearless* and HMS *Intrepid.* Of the other squadrons, No. 829 was responsible for operating Wessex helicopters from the Royal Fleet Auxiliary ships *Regent* and *Resource* while No. 846 reformed at Culdrose during July 1968, as the commando headquarters squadron. It was disbanded in December 1975, and reformed again in April 1976 as an operational commando squadron equipped with eight Wessex HU Mk 5s. It then carried out exercises in northern Norway operating from RFAs *Sir Galahad* and *Sir Tristam* and later from HMS *Hermes.*

Development of the Wessex HAS Mk 1 resulted in the HAS Mk 3 and three prototypes of this new variant were built, the first, XT255, flying on 30 November 1964. A number of HAS Mk 1s on the production line at Yeovil were completed as HAS Mk 3s, basically by the introduction of the 1600shp Gazelle Mk 165, a new search radar with the scanner housed in a large dorsal radome, a much improved dunking sonar, a new automatic flight control system and improvements to the navigational equipment. On 9 January 1967, No. 700H Flight, the Wessex Intensive Flying Trials Unit commanded by Lt. Cdr. D. R. V. Doe, reformed with five Wessex HAS Mk 3s at Culdrose. Deliveries of HAS Mk 3s to No. 814 Squadron started in August 1967 replacing its existing HAS Mk 1s and when No. 700H Flight disbanded in September 1967, it was integrated into No. 814 Squadron. For its first six months of operations the squadron provided small detachments to operate from aboard RFA ships. Then in May 1968 the whole squadron embarked in HMS *Hermes* for a period in the Far East. In October 1968, No. 826 Squadron re-equipped with six HAS Mk 3s for the ASW role and two HAS Mk 1s for service aboard HMS *Eagle* for the ASR role. By the time the squadron disbanded on 25 March 1970 it had spent three periods aboard *Eagle,* the first of which included a visit to the United States, while the other two were for exercises in the Mediterranean.

The HAS Mk 1s of No. 820 Squadron were replaced by four HAS Mk 3s in May 1969 to provide an ASW squadron for the helicopter cruiser HMS *Blake.* The squadron embarked in *Blake* on 30 June 1969 and was to spend much of the next two years aboard, on voyages to the Far East, the USA, and the Mediterranean. Only two other first line squadrons operated HAS Mk 3s and these were both headquarters squadrons which provided aircraft for operations on board a variety of ships. No. 819 provided Wessex helicopters for operation aboard RFA ships and No. 829 had the responsibility for the Wessex helicopters operating from the county class destroyers. The responsibility for these was transferred to No. 737 Squadron at Portland in June 1970. By December 1972 the HAS Mk 3 was obsolescent and had been taken out of first line service, but then it was discovered that its replacement, the Sea King, was too big for the hangars on the county class

destroyers so a small number of Wessex HAS Mk 3s have been retained on the strength of No. 737 Squadron for use on these ships.

Following the Argentinian invasion of the Falkland Islands on 2 April 1982, Britain launched "Operation Corporate" to recover the Falkland Islands, by initially assembling Task Force 317 to transport the British Force to the South Atlantic. Included in the Task Force were over 50 Wessex helicopters, virtually all HU Mk 5s but with two HAS Mk 3s of No. 737 Squadron on board the two county class destroyers HMS *Antrim* (100 Flight) and HMS *Glamorgan* (103 Flight). On 12 June 1982, *Glamorgan* was hit by an Exocet missile launched from the Falkland Islands, which fortunately did not explode although it set fire to the ship's hangar, causing serious damage to the hangar and destroying 103 Flight's Wessex, XM837. HMS *Antrim* was tasked with landing a small detachment of Special Forces on the Fortuna Glacier in South Georgia and for this purpose two of No. 845 Squadron's Wessex HU Mk 5s and HMS *Antrim*'s own Wessex HAS Mk 3 were successfully deployed. However, when the troops were being evacuated the following day both of 845 Squadron's HU Mk 5s crashed in 'white-out' conditions, fortunately with no fatalities. HMS *Antrim*'s HAS Mk 3, XP142, successfully returned to the ship and, when the weather had improved, rescued the remainder of the force from the glacier.

Three days later XP142, which was nicknamed "Humphrey", was in action again, dropping depth charges on the Argentine submarine *Santa Fe* just outside Grytviken harbour, while it was trying to make its escape to the open sea. This attack, along with subsequent attacks by other Fleet Air Arm helicopters, forced the *Santa Fe* to return to Grytviken where it was beached.

In addition to the two HU Mk 5s lost on the Fortuna Glacier and the HAS Mk 3 destroyed by fire in HMS *Glamorgan,* a further six HU Mk 5s, all belonging to No. 848 Squadron, were destroyed aboard the *Atlantic Conveyor* when the ship was set on fire on 25 May 1982 after being hit by an Exocet missile launched from a Super Etendard. Three squadrons equipped with HU Mk 5s took part in Operation Corporate and of these only No. 845 Squadron was operational before the Argentinian invasion, the other two, No. 847 and No. 848, being reformed specifically to take part in the operation. Each of the squadrons was split into a number of flights to operate from various ships of the Task Force. The Wessex helicopters were used very successfully for a wide variety of duties ranging from the attack role armed with missiles to the casualty evacuation (CASEVAC) and search and rescue (SAR) roles. By the end of 1982 both No. 847 and No. 848 Squadrons had disbanded leaving No. 845 as the last remaining first line Wessex squadron. Eventually this squadron converted to Sea King HC Mk 4s on 31 October 1986, leaving just a few Wessex HU Mk 5s in service, primarily in the SAR role until the type was withdrawn from service with the Fleet Air Arm on 31 March 1988.

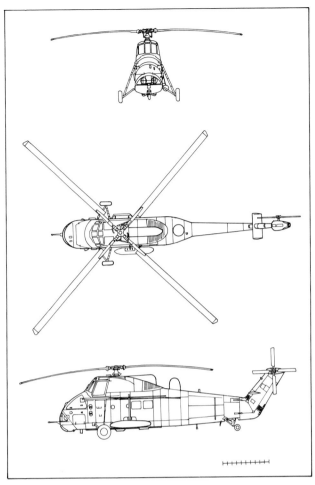

33   Wessex HAS Mk 3

PRODUCTION

Prototype (Sikorsky built)   XL722.
HAS Mk 1 XL727 – XL729 (Prototypes).
HAS Mk 1/3   XM299 – XM301, XM326 – XM331,
XM832 – XM845, XM868 – XM876, XM915 – XM931,
XP103 – XP118, XP137 – XP160, XS115 – XS128,
XS149 – XS154, XS862 – XS889.
HAS Mk 3 XT255 – XT257 (Prototypes).
HU Mk 5 XS241 (Prototypes), XS479 – XS500,
XS506 – XS523, XT448 – XT487, XT755 – XT774.

CHAPTER THIRTY-FOUR

De Havilland Sea Heron

Following the initial success of the DH104 Dove, de Havilland's considered that there was an opening in the market for a 14 to 17 seat feeder airliner. This requirement they believed could be met by an enlarged, four engined development of the Dove. Design work on this new project, the DH114, started in 1949, and the aim was to keep the design simple, where possible, using existing Dove components. The DH114 had a fixed tricycle undercarriage and was powered by four ungeared, unsupercharged de Havilland Gipsy Queen 30 engines driving two-bladed, de Havilland, variable pitch propellers.

The first flight of G-ALZL, the prototype DH114, which by then had been named the Heron, took place at Hatfield on 10 May 1950 with G. H. Pike at the controls. The early test flights identified a control

problem which was quickly corrected by giving the tailplane some dihedral. Except for this one problem the test flight programme proceeded exceedingly well and within a year the aircraft had been issued with a full Certificate of Airworthiness and deliveries of production aircraft started early in 1952.

At an early stage in the Heron's development, a Series 2 version with a retractable undercarriage was produced and the prototype Heron 2, G-AMTS, was the seventh production aircraft and also the last to be built at Hatfield because at that time the production line was in the process of being transferred to the de Havilland factory at Chester. G-AMTS flew for the first time on 14 December 1952. The retractable undercarriage resulted in a significant reduction of drag which gave the aircraft increases in performance and fuel economy. The Heron subsequently sold quite well both as a feeder airliner and as an executive transport

Sea Heron C Mk 2, XR441, of Yeovilton Station Flight. (HMS *Heron*)

Sea Heron C Mk 2, XR444, of No. 781 Squadron, Lee-on-Solent.

with a total of 148 being built by the time production ended in 1961.

During 1961 five civil Herons were bought by the Royal Navy to supplement its Sea Devon C Mk 20s. It was considered that the larger size of the Heron would make it rather more versatile than the Devon, but still very economical to operate. Two of the aircraft, G-AORG and G-AORH, came from Jersey Airlines, and received serial numbers XR441 and XR442 respectively. The other three, from West African Airways, were VR-NAQ, VR-NCE and VR-NCF and received serial numbers XR443, XR444 and XR445 respectively. These aircraft, identified as the Sea Heron C Mk 2, were delivered to Lee-on-Solent in July 1961 where they joined No. 781 Squadron, the Fleet Air Arm's Communications Squadron commanded by Lt. Cdr. R. C. Stock. For a while No. 781 Squadron also operated two Herons which had previously belonged to the RAF's Queen's Flight. Sea Herons were still on the strength of No. 781 Squadron when it disbanded in March 1981 as part of an economy drive. No. 728 Squadron, the Fleet Requirement Unit at Hal Far in Malta, operated Sea Heron XR444 from March 1963 until October 1965 and during the final year it also acted as the Hal Far Station Flight in addition to its normal squadron duties. Rather appropriately the Station Flight at RNAS Yeovilton – HMS *Heron* – has operated Sea Herons since November 1972 and since the disbandment of No. 781 Squadron has become the principal operator of the Sea Heron. In addition to the normal communications duties the aircraft are used for search and rescue (SAR) duties and fishery protection patrols.

Since they entered service in 1961 the Sea Herons have given excellent service and now, some 27 years later, four of the original five are still in regular use, and the quantity has been made up to five by the continued use of the ex RAF Heron CC Mk 4 XM296. The Sea Herons are expected to remain in service for some time to come.

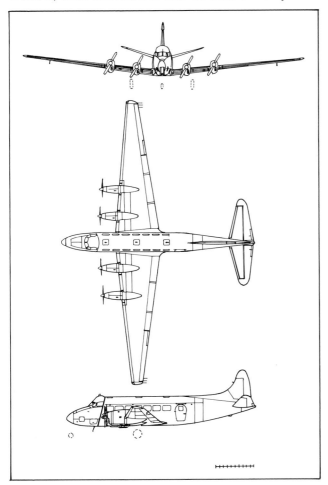

34   Sea Heron C Mk 2

PRODUCTION

C Mk 2   XR441 – XR445.

CHAPTER THIRTY-FIVE
Westland Wasp

Shortly after the Second World War the Cierva Autogiro Company had designed and built a light, two-seat helicopter trainer known as the W.14 Skeeter. The first prototype registered G-AJCJ flew for the first time on 8 October 1948 at Eastleigh, with an improved Mk 2 version G-ALUF being completed towards the end of 1949. Although neither of these prototypes was good enough to achieve production status the Ministry of Supply was sufficiently impressed with the project to place an order for two prototypes of the Skeeter Mk 3 which were to be built to Specification A.13/49 for

The third prototype P.531-O/N, XN334/758, of No. 771 Squadron, carrying out deck trials on board HMS *Undaunted*, after the fitting of suction pads to the skids. (Westland)

evaluation and trials purposes. However before these had been completed the Cierva Company, which was having financial problems, was taken over by Saunders-Roe Limited.

The two Skeeter Mk 3s, WF112 and WF113, were completed by Saunders-Roe and were used for evaluation and service trials including a period with the Royal Navy's helicopter training squadron, No. 705 at RNAS Gosport. The performance of the Skeeter was considered to be satisfactory, although like its predecessors the Mk 3 suffered from a ground resonance problem that the manufacturers seemed unable to resolve. To meet a Royal Navy requirement for a light observation helicopter, capable of operating from small

Wasp HAS Mk 1, XT779/452, of No. 829 Squadron, Galatea Flight. (RNAS Culdrose)

ships, the design was further refined, and an order was placed for a single Skeeter Mk 4, WF114. Again, this aircraft suffered from ground resonance problems and proved to be totally unsatisfactory. However WF114 was used for some intensive experimental work by the company to cure the resonance problem with apparent success as the next version, the Skeeter Mk 5, proved to be clear of the problem. By this time the Royal Navy had lost interest in the Skeeter, although the Army retained an interest and ordered 64 for use as air observation posts in addition to the RAF ordering three as dual control trainers. A small number were also exported for use by the West German Army and Navy.

In 1956 Saunders-Roe came to the conclusion that there was an increasing need for a light five-seat helicopter for both the Army and Royal Navy and immediately set to work designing a helicopter to this new requirement, using as many existing Skeeter components as possible, and to be powered by a 325shp Blackburn Turmo turbine engine. Two of these heli-

Wasp HAS Mk 1, XV636/606, of No. 829 Squadron, launching an AS-12 wire guided missile. (HMS Osprey)

copters, identified as the Type P531, were built as a private venture; the first G-APNU flying on 20 July 1958 and the second G-APNV on 30 September 1958. Two virtually separate lines of development were then taken, leading to the Scout used by the Army and the Wasp used by the Royal Navy.

The second prototype, G-APNV, was allocated serial number XN332 in 1959 for trials by the Royal Navy becoming in effect the first prototype Wasp, and was followed shortly afterwards by two further prototypes, XN333 and XN334; all three prototypes being designated the type P531-O/N. As the intention was to operate these helicopters from platforms on small ships, mainly frigates, a great deal of effort was concentrated on developing a suitable landing gear. Several alternatives were considered including skids, suckers and a long stroke quadricycle wheeled undercarriage. To simplify this work a platform was built at RAE Bedford which could be moved to simulate a pitching deck of a frigate in a stormy sea. Following the trials it was decided to standardise on a four point castored wheel undercarriage for the production aircraft. Trials using the P531-O/Ns were carried out by No. 700X Flight and No. 771 Squadron. Following the successful completion of these trials in 1961 two pre-production aircraft, XS463 and XS476, were ordered

Buccaneer S Mk 1, XN968/105/E, of No. 800 Squadron on HMS *Eagle*, 1964. (B. J. Lowe)

programme requirements they transferred their flight test facilities to the wartime bomber airfield at Holme-on-Spalding Moor. On 9 July 1958 the first prototype, XK486, was flown from Bedford to Holme-on-Spalding Moor, giving a display to the workers at Brough en route. The second prototype, XK487, flew for the first time on 26 August 1958 from the test airfield and shortly afterwards along with XK486 was flown to Farnborough where it was put on static display for the 1958 SBAC Show. The first prototype having by then achieved some 25 flying hours was demonstrated in the flying display by D. J. Whitehead. During the show the first production contract for 40 aircraft was announced.

Shortly after its first flight on 3 October 1958 the third prototype, XK488, was delivered to the de Havilland Engine Company at Hatfield for development of the D.H. Gyron Junior engine. The fourth prototype, XK489, was the first of the navalised NA39s having folding wings and an arrester hook. The NA39 was also too long for the carrier lifts and consequently had to have a hinged nose section which could be readily folded back. This aircraft was initially delivered to the A&AEE, Boscombe Down for trials including tropical trials in Malta during the summer of 1959. The next aircraft, XK490, was the first to be fitted with a rotating

Buccaneer S Mk 1, XN928/119/R, of No. 801 Squadron, HMS *Ark Royal*, September 1963. (RAF Museum)

bomb bay and door and consequently was allotted for armament trials. After being demonstrated at the Paris Salon in June 1959 and the SBAC Show at Farnborough in September, XK490 was destroyed with the loss of the A&AEE crew on 12 October 1959 when flying on trials from Boscombe Down. The remaining prototypes were used for a variety of trials including flight refuelling, and weapon and avionic systems.

Initial deck landing trials were carried out during January 1960 on HMS *Victorious* using two NA39s, XK489 and XK523, the first landing and take-off being made by Blackburn's chief test pilot D. J. Whitehead. In the four days of the trials some 30 take-offs and landings were successfully achieved.

On 26 August 1960 the name Buccaneer was officially adopted for the NA39 and the aircraft of the first production order were identified as the Buccaneer S Mk 1. Shortly after this, following its appearance at the 1960 SBAC Show at Farnborough, the first prototype, XK486, crashed on 5 October 1960, the pilot, G. R. I. "Sailor" Parker, and his observer successfully ejecting from the stricken aircraft. Sadly "Sailor" Parker and his flight observer G. R. C. Copeman were later killed when XN952 crashed at Holme-on-Spalding Moor on 19 February 1963.

Four of the pre-production aircraft, XK526 to XK529 inclusive, were delivered to the Royal Naval Test Unit at the A&AEE, Boscombe Down for the continuation of carrier trials. The first two, XK526 and XK527, were used for trials on HMS *Ark Royal* during a Mediterranean cruise early in 1961, the former being subsequently used for tropical trials in Singapore. On 31 August 1961, during trials on HMS *Hermes* in Lyme Bay, XK529 crashed into the sea with the loss of its Royal Navy crew.

The last six pre-production Buccaneers, XK531-XK536, were used by No. 700Z Flight, the Buccaneer Intensive Flying Trials Unit (IFTU), which was commissioned at RNAS Lossiemouth on 7 March 1961 under the command of Cdr. A. J. "Spiv" Leahy. The first two aircraft arrived in May 1961 and by the end of that year the unit was up to full strength although one aircraft, XK535, was destroyed in a landing accident at Lossiemouth on 18 August 1962. During 1962 the IFTU formed an aerobatic team using four Buccaneers which gave a spirited display at the SBAC Show at Farnborough that September. By the end of 1962 the IFTU had completed its programme of trials and on 15

January 1963 No. 700Z Flight was disbanded.

On 17 July 1962, No. 801 Squadron, commanded by Lt. Cdr. E. R. Anson, was commissioned at Lossiemouth with eight of the early production Buccaneer S Mk 1s. After a period of working-up, the squadron embarked in HMS *Ark Royal,* operating in the English Channel, on 19 February 1963 and after an intensive period of carrier-borne operations the squadron returned to Lossiemouth. After increasing its strength to ten Buccaneer S Mk 1s No. 801 Squadron embarked in HMS *Victorious* on 14 August 1963 for operations East of Suez. Following a period in Malayan waters *Victorious* sailed to Dar-es-Salaam to relieve HMS *Centaur,* disembarking her Buccaneers to Embakasi on 7 February 1964 from where they were to provide support if required for the ground forces dealing with a rebellion in East Africa. The squadron re-embarked two weeks later and *Victorious* returned to the Far East where she was to remain until June 1965 when she sailed for the United Kingdom. On arrival in British waters the squadron was flown off to Lossiemouth where it was disbanded on 27 July 1965.

A new squadron, No. 809, commanded by Lt. Cdr. J. F. H. C. de Winton, was commissioned in January 1963 at Lossiemouth, using the aircraft and crews of the recently disbanded No. 700Z Flight. This squadron subsequently expanded to 14 aircraft and in addition to maintaining itself as an operational squadron it had to continue with various operational development trials and also operate as the Buccaneer Operational Flying Training Squadron. With the completion of the Buccaneer S Mk 1 programme the squadron was disbanded on 26 March 1965. The operational flying training aspect of their work was reduced to second line status and became the responsibility of No. 736 Squadron, using ex No. 809 Squadron aircraft and crews at Lossiemouth. This squadron continued to operate Buccaneer S Mk 1s until December 1970, long after they had been phased out of first line service.

The third operational squadron equipped with Buccaneer S Mk 1s was No. 800 commanded by Lt. Cdr. J. C. Mather and reformed on 18 March 1964 at Lossiemouth. In support of the squadron, No. 800B Flight, equipped with four Scimitar F Mk 1s, was formed on 9 September 1964. These Scimitars were to act as "buddy"

tankers for the Buccaneer S Mk 1s both at Lossiemouth and when embarked in HMS *Eagle.* Due to weight limitations on take-off the Buccaneers could be launched with only a limited amount of fuel and would, if necessary, be topped up by the Scimitars immediately after take-off.

Whilst working up at Lossiemouth No. 800 Squadron took part in the Fleet Air Arm Jubilee Review at Yeovilton on 28 May 1964 and then together with No. 800B Flight, embarked on 2 December 1964 aboard HMS *Eagle* which, after a period in Home waters, sailed for the Indian Ocean. There No. 800 Squadron's Buccaneers were to see operations against rebel tribesmen in Aden as well as carrying out patrols to enforce the government imposed oil embargo on Rhodesia. By the end of 1966 No. 800 Squadron's Buccaneer S Mk 1s had been replaced by the much improved Buccaneer S Mk 2s.

It had soon become apparent during development that the Buccaneer S Mk 1 was underpowered and the loss of an engine during a maximum weight take-off would almost certainly have been disastrous. In addition the Gyron Junior had proved to be an uneconomical engine giving the Buccaneer a very poor radius of operation. Fortunately Rolls Royce had started development of the Spey, a new advanced turbo-fan engine, for the de Havilland DH121 Trident medium range airliner which had been ordered by British European Airways. The Spey met all the basic requirements for a Gyron Junior replacement, producing some 40% more thrust with a substantially lower fuel consumption and was of a size that could be accommodated in the Buccaneer. The main modification to the airframe was the enlargement of the air intakes to cater for the increased mass air flow required for the Spey. In addition boundary layer control slits were introduced in the inner wing to improve landing performance.

A production order for 84 Buccaneer S Mk 2s was placed on 8 January 1962 and to expedite the development programme two of the original development aircraft, XK526 and XK527, were returned to Blackburn's for conversion into prototypes for the S Mk 2. The first of these, XK526, flew for the first time on 17 May 1963, followed by the second, XK527, on 19 August 1963. The first production Buccaneer S Mk 2, XN974, flew on 6 June 1964 and together with the next two production aircraft, XN975 and XN976, joined the two prototypes on the development programme. Of

Prototype Buccaneer S Mk 2, XK526, converted from one of the original Mk 1 development batch aircraft. (B.Ae.)

Buccaneer S Mk 2, XV868/020/R, of No. 809 Squadron, ready for launch on the bow catapult on HMS *Ark Royal*. (Lt. Cdr. M. S. Lay)

these XN975, flown by D. J. Whitehead and four Royal Navy pilots, carried out the carrier trials on board HMS *Ark Royal*, achieving some 78 launches in seven days. XN976 appeared at the SBAC Show, Farnborough, in September 1964, fitted with vortex generators at the wingtips and carrying two rocket pods and two Martin Bullpup air-to-surface missiles under the wings. For the A&AEE tropical trials three aircraft, XK527, XN974 and XN976, were flown to the US Navy base at Pensacola. As part of these trials the aircraft operated for four days aboard the USS *Lexington*, carrying out some 100 take-offs. On completion of the trials Cdr. G. R. Higgs AFC, the commanding officer of the Naval Test Squadron at the A&AEE with Lt. Cdr. A. Taylor as observer, flew direct from Goose Bay in Labrador to Lossiemouth, covering the distance of 1950 miles in 4 hours 16 minutes. The last of the development aircraft, XN893, was flown to Hucknall on 23 February 1965 where it was used by Rolls Royce for engine development.

Service trials with the Buccaneer S Mk 2 started with the commissioning of a new IFTU, No. 700B Flight commanded by Cdr. N. J. P. Mills, at Lossiemouth on 9 April 1965, which was ultimately to operate a flight of nine aircraft. The trials were soon completed and the unit disbanded on 30 September 1965.

The first operational squadron to be equipped with the Buccaneer S Mk 2 was No. 801 commanded by Lt. Cdr. J. F. H. C. de Winton, which reformed at Lossiemouth with 12 aircraft on 14 October 1965: the aircraft and crews of the recently disbanded No. 700B Flight were being transferred to form the nucleus of the squadron. On 21 October 1965, the 160th Anniversary of Trafalgar Day, the squadron's Commanding Officer flew over Nelson's Column at 1000ft in XN980/233. After completing its working-up at Lossiemouth the

squadron embarked in HMS *Victorious* in May 1966 for three weeks at sea to complete its training. On 8 July 1966 the squadron re-embarked in *Victorious* and was to spend the next year in the Far East, where it took part in joint exercises with the USS *Enterprise*. In 1967 No. 801 Squadron was awarded the Boyd Trophy for its involvement in successfully bringing the Buccaneer S Mk 2 into operational service. The squadron embarked in HMS *Hermes* for a further spell of nine months in the Far East and after a final cruise on *Hermes* in the Mediterranean during 1970, was disbanded on 21 July 1970.

The second squadron with Buccaneer S Mk 2s was the reformed No. 809 commanded by Lt. Cdr. L. E. Middleton, which commissioned at Lossiemouth on 27 January 1966 and took part in a flypast of Fleet Air Arm aircraft at the 1966 SBAC Show at Farnborough. Following a three week spell embarked in HMS *Victorious* during May 1966, the squadron was to operate from *Hermes* until May 1970 when it became part of HMS *Ark Royal*'s air group and was to remain so until, with the impending withdrawal of *Ark Royal* from service, the squadron was flown off for the last time on 27 November 1978. Flying direct to the RAF Maintenance Unit at St. Athan, where the aircraft were transferred to the RAF and the squadron officially disbanded at Lee-on-Solent on 15 December 1978.

Buccaneer S Mk 2s began to replace the Buccaneer S Mk 1s of No. 800 Squadron at Lossiemouth in June 1966 but it was not until November 1966 that the replacement was complete. During March of the following year the oil tanker *Torrey Canyon* ran aground on the Seven Stones Reef near Land's End and started to leak thousands of tons of crude oil into the sea, putting the Cornish holiday beaches seriously at risk from pollution. In an attempt to minimise the pollution, the Buccaneers of No. 800 Squadron along with those of No. 736 Squadron were ordered to destroy the tanker and its cargo. Flying from RNAS Brawdy on 28 March

1967 eight Buccaneers dropped 42000lb of high-explosive bombs and achieved a 75% success rate. The squadron provided the strike capability for HMS *Eagle* during its operations in Home waters, the Mediterranean and the Far East until *Eagle* was prematurely withdrawn from service in 1971 and No. 800 Squadron was disbanded at Lossiemouth on 23 February 1972.

The fourth and last Buccaneer S Mk 2 Squadron was No. 803 which, under the command of Lt. Cdr. M. J. A. Hornblower, initially reformed at Lossiemouth on 3 July 1967 and began equipping with Buccaneer S Mk 2s in January 1968, completing the replacement programme by August 1968. This squadron operated at Lossiemouth as the Buccaneer Headquarters Squadron carrying out a series of weapons trials and from 23 August until 1 April 1969 a detachment of four aircraft operated from HMS *Hermes* in the Indian Ocean. The squadron finally disbanded on 18 December 1969.

In the mid-1970s the Martel air-to-surface missile was introduced into Fleet Air Arm service and resulted in the Buccaneers being re-designated the S Mk 2C for non-Martel aircraft and the S Mk 2D for the aircraft being equipped to carry the Martel. No. 809 was the only Fleet Air Arm squadron still operating Buccaneers by that time and operated a mixed squadron of S Mk 2Cs and S Mk 2Ds from October 1973 until it disbanded in December 1978.

From the early days of Buccaneer development the Blackburn Aircraft Company were confident that it would achieve some overseas sales. Initially there were hopes that they would sell it to the US Navy but in the event the Grumman A6 Intruder was selected. Starting in 1960 a campaign was launched to sell the Buccaneer

S Mk 1 to West Germany. Demonstrations for the Germans were held both in Germany and Britain but all these efforts came to nothing. Attempts to sell the Buccaneer to South Africa proved to be more successful and in October 1962 the South African Government placed an order for 16 aircraft and followed this with an option for a further 14. The South African aircraft, identified as the Buccaneer S Mk 50, were similar to the S Mk 2 but without the naval equipment and with Bristol Siddeley BS605 rocket motors fitted to improve take-off performance on the hot high airfields that are prevalent in South Africa. One of the 16 Buccaneers was lost on the delivery flight so No. 24 Squadron South African Air Force was formed with 15 Buccaneers. South African attempts to exercise its option for a further 14 aircraft for its airforce have been continually frustrated by the British Government and no further Buccaneers were delivered to South Africa.

The only other operator of the Buccaneers was the RAF who came very late on to the Buccaneer scene. Following the cancellation of the TSR-2 in 1965 and the General Dynamics F.111 in 1968, the RAF was left without any projected strike aircraft, leaving the field wide open for the Buccaneer, which by then had been fully developed, and already proved to be an excellent strike aircraft by the Fleet Air Arm and South African Air Force. Initially the government would only authorise a contract for 26 Buccaneers and as the decision had also been made to run down the fixed wing element of the Fleet Air Arm it was decided that 62 of their 84 Buccaneers were to be transferred to the RAF and with an absolute minimum of modifications. The ex Fleet Air Arm Buccaneers were identified as the S Mk 2A and the new build, which were to full RAF standard, the S Mk 2B. The main differences were the removal of the equipment for ship-borne operation and the alteration of the communication and avionic fit to bring it into line with RAF practice. Subsequently, a further order for 25 Buccaneer S Mk 2Bs was placed. By 1980 the RAF were operating five squadrons including Nos. 15 and 16 Squadrons in Germany, but with the advent of the British Aerospace Tornado GR Mk 1 these two squadrons were disbanded in 1983 and 1984 respectively. The other three squadrons operating in the UK were Nos. 12, 208 and 216. In 1980 No. 216 merged with No. 12 leaving two squadrons, Nos. 12 and 208, operating from RAF Lossiemouth in the maritime strike role. It is expected that these two squadrons will continue to operate the Buccaneer for many years to come.

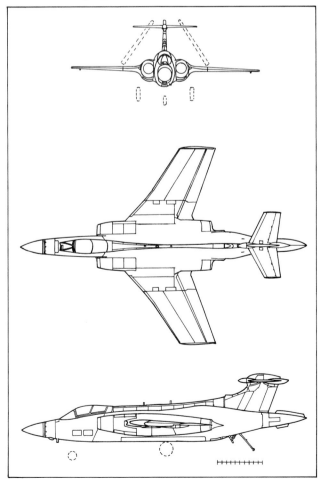

36   Buccaneer S Mk 1

PRODUCTION

Prototypes   XK486, XK487.
Pre-production   XK488 – XK491, XK523 – XK536.
S Mk 1   XN922 – XN935, XN948 – XN973.
S Mk 2   XN974 – XN983, XT269 – XT288, XV152 – XV168, XV332 – XV361, XV863 – XV869.

CHAPTER THIRTY-SEVEN
De Havilland Chipmunk

The DHC 1 Chipmunk was the first aircraft to be designed and built by the Toronto based de Havilland Aircraft of Canada Ltd and was a basic trainer intended as a Tiger Moth replacement. The Chipmunk was a low-wing monoplane of conventional stressed skin construction but with fabric covered flying control surfaces. The pupil pilot and instructor were seated in tandem under a long sliding canopy. The prototype flew for the first time at Toronto on 22 May 1946 in the hands of W. P. I. Fillingham the chief production test pilot at the de Havilland, Hatfield factory. Production was soon under way and by 1951 when production ceased temporarily in Canada, a total of 158 aircraft had been built, mainly for the Royal Canadian Air Force (RCAF). An additional 60 Chipmunks were built in Canada in 1956 for the RCAF.

Two of the early production Chipmunks were delivered to de Havilland at Hatfield where they were equipped to British standard and given British civil registrations G-AJVD and G-AKDN. These two aircraft were used by the company for trials and were then sent in November 1948 for Service Trials to the A&AEE at Boscombe Down. Following minor modifi-

cations to the undercarriage and elevator the aircraft was selected as the standard Ab Initio trainer for the RAF and ordered into large scale production.

Production of the British Chipmunks started at Hatfield, but after 111 had been built the production line was transferred to the de Havilland factory at Chester, where a further 889 were built. Of the 1000 Chipmunks built in Britain, 740 were Chipmunk T Mk 10s for the RAF; although a small number of the remaining 260 went to civil customers, the majority were delivered to overseas air forces.

The RAF's Chipmunk T Mk 10s were designed and built to meet Specification T.8/48 and they first entered service early in 1950 when they started to replace the Tiger Moths of the Oxford University Air Squadron. By the time the last one had been delivered to the RAF in 1953 all the university air squadrons had been re-equipped with Chipmunks, as had many of the RAF and RAFVR training units.

Overseas orders kept Chipmunk production going at Chester until February 1956 when the last production aircraft was delivered to the Royal Saudi Arabian Air Force. In the mid 1950s, with the closure of the RAFVR Flying Schools and the RAF move towards all jet flying training, a large number of the RAF's

Chipmunk T Mk 10, WP906/816, of No. 771 Squadron, Culdrose. (RNAS Culdrose)

Chipmunk T Mk 10, WP809/912, of the Britannia Flight, Roborough, which provides air experience for the students of the Royal Naval College, Dartmouth. (FONAC)

Chipmunks became surplus and many were sold to civilian operators.

The Royal Navy's association with the Chipmunk began in 1965 when it obtained twelve Chipmunk T Mk 10s from the RAF to replace the Tiger Moths which it still had in service. Since that initial batch of twelve aircraft a further four have been taken on charge. The Britannia Royal Naval College Air Experience Flight was the first Royal Navy unit to operate Chipmunks, receiving their aircraft in June 1966 and remaining, up to the time of writing, the prime Royal Navy operator of the type. Only two squadrons, both second line, have used Chipmunks and these are No. 781, the Southern Communications Squadron based at Lee-on-Solent, which used a single Chipmunk for glider towing and general duties, from July 1971 until disbanded in March 1981. The other squadron, No. 771 based at RNAS Culdrose, currently operates two Chipmunks received in January 1983 on station flight duties. The only other Royal Navy units to use Chipmunks are: the Station Flights at RNAS Yeovilton from June 1971 until the present; RNAS Lossiemouth from October 1971 until January 1972, and RNAS Culdrose from March 1972 until October 1974. The small number of Chipmunks operated by the Royal Navy have given good service and as there are no plans for a replacement are likely to continue to do so for some time to come.

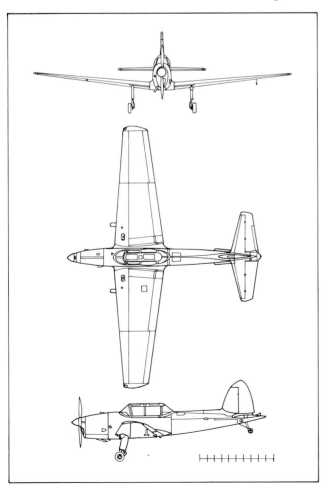

37   Chipmunk T Mk 10

PRODUCTION

T Mk 10 WB575, WB657, WB671, WD374, WK511, WK574, WK608, WK634, WK635, WP776, WP795, WP801, WP809, WP856, WP904, WP906.

Phantom FG Mk 1, of No. 892 Squadron, after being prepared for launch on HMS *Ark Royal*'s bow catapult. (Lt. Cdr. M. S. Lay)

J79-2 engines, were identified as the F-4A and these were followed by the J79-8 powered F-4B of which 649 were built for the US Navy and Marine Corps. A small number of RF-4Bs, a reconnaissance version of the Phantom, were built for the US Marine Corps. The final version built for the US Navy and Marine Corps was the F-4J which saw the introduction of 17900lb thrust J79-10 engines and various improvements to the radar and weapons systems. The first F-4J flew for the first time in May 1966 and when production came to an end in 1972 a total of 522 had been delivered. Towards the end of the 1970s, 265 F-4Js were modernised by improvements to the structure and the introduction of the AWG-10A weapon control system and the improved J79-108 engines.

Meanwhile in Britain the Hawker P1154 project development programme was under way and it soon became apparent that there would be problems in attempts to produce an aircraft to meet the widely differing RAF and RN requirements. The RAF required a single-seat strike aircraft equipped with terrain-following radar and capable of supersonic speed at low level while the RN requirement was for a two-seat, all-weather fighter with a long endurance and capable of extended supersonic flight at high altitude. Towards the end of 1963, McDonnell noted the Royal Navy's lack of enthusiasm for the P1154 and immediately re-opened its sales campaign, this time with strong support from Rolls Royce. Early in 1964, the Naval version of the P1154 was cancelled and on 1 July government approval was given for the procurement of the F-4 Phantom for the Royal Navy. An order was placed immediately for two Spey engined YF-4K pre-production aircraft for development purposes and shortly afterwards was increased to provide two further aircraft for systems development.

McDonnell's initial intention for the Royal Navy's F-4K Phantom was a relatively simple development of the US Navy's F-4B, fitted with Rolls Royce Spey engines, Martin Baker ejection seats and other radar and communications systems that would be appropriate for Fleet Air Arm operations. However, in the meantime, a later version of the Phantom, the F-4J, would be used as the basis of the F-4K. The initial assumptions that the introduction of the Spey in place of the J-79 would be relatively simple proved to be false and a major redesign of the fuselage centre section was necessary to accommodate the larger diameter Speys. In addition to suit operation from the smaller carriers operated by the Royal Navy it was necessary to introduce high lift devices which included drooped ailerons, larger leading edge flaps, boundary layer control and an extra-extensible nose undercarriage leg to increase the angle-of-attack for take-off.

The first YF-4K Phantom, XT595, flew on 27 June 1966 followed on 30 August 1966 by the second, XT596, and although powered by two Rolls Royce Spey Mk 201 engines they were not fully representative of the production F-4K as they were still fitted with mainly US equipment which was due to be replaced by British manufactured equipment in later aircraft. These first two aircraft were allotted to the engine development flying programme. The third and fourth aircraft,

XT597 and XT598, were equipped to production standard and having flown for the first time on 1 November 1966 and 21 March 1967 respectively, were used for systems and weapons trials. During 1966 a production contract was placed for 18 F-4Ks and a further 39 were ordered the following year.

Trials showed the F-4K to have a better take-off and low altitude performance than the equivalent US versions but the changes in the fuselage to accommodate the Speys increased the drag which negated many of the benefits obtained from this engine, particularly at high altitude. During the development programme, problems were encountered with the handling of the new military Spey engine with reheat. These were to delay the Phantom deliveries and were not completely resolved until after it had entered service with the Fleet Air Arm. Carrier trials were successfully carried out on the USS *Coral Sea* in July 1968 using XT597, with XT857 as a back-up aircraft.

The Fleet Air Arm's first three Phantoms, FG Mk 1s, XT858, XT859 and XT860, were delivered from St. Louis to RNAS Yeovilton on 29 April 1968. The following day the Phantom Intensive Flying Trials Unit, commanded by Cdr. A. M. G. Pearson, was formed at Yeovilton, equipped initially with the three newly delivered Phantoms and progressively building up to its full strength of seven aircraft, which was achieved on 30 October 1968 with the arrival of XT869. It fell to No. 700P Flight to provide the Phantom FG Mk 1 demonstration aircraft for the SBAC Show at Farnborough in September 1968 and XT859/725/VL was used to provide a spirited display during which it produced several sonic booms. Having completed the trials, No. 700P Flight was disbanded on 31 March 1969

Two Phantom FG Mk 1s, XT872/004/R and XT859/006/R, of No. 892 Squadron, with the special Silver Jubilee markings on the nose.

to become the nucleus of No. 892 Squadron which was to be the only Fleet Air Arm first line squadron to be equipped with the Phantom. However before this, on 14 January, No. 767 Squadron under the command of Lt. Cdr. P. G. Marshall had reformed at Yeovilton as the Fleet Air Arm's Air Warfare Fighter Training (Phantom) Unit and was responsible for the conversion training of pilots and observers for the Phantom FG Mk 1.

Following the government decision not to proceed with the new aircraft carrier, CVA-01, intended to replace HMS *Ark Royal* in the late 1970s, there was no longer a requirement for the planned second front line Phantom squadron and consequently a number of the Phantom FG Mk 1s ordered for the Royal Navy were transferred to the RAF, which used them to equip No. 43 Squadron at RAF Leuchars. Training of RAF crews was carried out by No. 767 Squadron, which provided all the operational training for both No. 892 Squadron and No. 43 Squadron until it disbanded on 1 August 1972, because of a policy decision by the Ministry of Defence that all Fleet Air Arm fixed wing flying training would in future become the responsibility of the RAF.

To replace No. 767 Squadron the Post Operational Conversion Unit Phantom Training Flight, thankfully later renamed the Phantom Training Flight (PTF), was formed at RAF Leuchars with Fleet Air Arm instructors, although under the command of an RAF officer. With the, apparently, very limited prospects for fixed wing aircrew in the Fleet Air Arm, there was an exodus of officers out of the service and to make up the numbers required to keep No. 892 Squadron operational, a number of RAF aircrew were seconded to the Royal Navy. The Phantoms transferred from No. 767 Squadron to the PTF retained their Fleet Air Arm colour scheme, although the legend "ROYAL NAVY"

Phantom FG Mk 1, XT872/001/R, of No. 892 Squadron, ready for launch from the bow catapult, as signified by the raised flag of the Flight Deck Officer (FDO). (Lt. Cdr. M. S. Lay)

was deleted from above the serial number on the fuselage sides. The PTF disbanded on 31 May 1978 shortly after HMS *Ark Royal* had set out on her final cruise and the aircraft, still in the Royal Navy colour scheme, were repainted in the contemporary RAF scheme.

No. 892 Squadron, the Fleet Air Arm's only first line Phantom FG Mk 1 Squadron, reformed at RNAS Yeovilton on 31 March 1969 under the command of Lt. Cdr. B. Davies, AFC, with a nucleus of experienced crews from the IFTU, No. 700P Flight. Shortly after reforming it was announced that the squadron was entering three crews in the Daily Mail Transatlantic Air Race which was to be held between the 4-11 May 1969 to commemorate the 50th Anniversary of the first transatlantic flight. The race was to be from the top of the Empire State Building in New York to the top of the Post Office Tower in London.

The planning of the race proved to be a major feat of organisation, optimising the flight plan for maximum speed and fuel consumption and ensuring that the Victor K Mk 1A tankers from No. 55 Squadron RAF would be in the right place at the right time. It was the aircraft's observers that were officially entered in the race. They had to get from the top of the Empire State Building and then by motorcycle and helicopter to the Floyd Bennett Naval Air Station, where the Phantoms would be waiting to take off. During the flight the aircraft would be refuelled three times by Victor tankers located at appropriate points on the route and then, after landing at the British Aircraft Corporation airfield at Wisley, the observers would be flown to the Post Office Tower by a Wessex helicopter.

On 4 May 1969, XT860/002/R flown by Lt. Cdr. D. Borrowman, with Lt. P. Waterhouse as the observer, completed the race in 5 hrs 31 mins, the flight taking 5 hrs 3 mins. The second crew, comprising Lt. A. Hickling and Lt. M. Drake flying in XT861/003/R, on 7 May 1969 improved on the first with a total time of 5 hrs 19

mins. The third crew, comprising the squadron C.O., Lt. Cdr. B. Davies and senior observer, Lt. Cdr. P. Goddard, flying XT858/001/R on the last day of the race, set a new record of 5 hrs 11 mins: this included a record New York to London point to point time of 4 hrs 47 mins, becoming the overall winner of the race and being presented with a prize of £6,000.

The squadron continued to work up but was unable to complete its training as HMS *Ark Royal,* which was the only Royal Navy carrier capable of operating the Phantoms, had not completed her refit and would not be available until mid 1970. To allow the completion of its training programme, arrangements were made for the squadron to embark in the USS *Saratoga* (CV-60), then operating in the Mediterranean as part of the US Sixth Fleet, for a period of deck landing training from 18 to 24 October 1969.

Following the major refit at Devonport Dockyard, HMS *Ark Royal* was recommissioned in February 1970 and on 30 April a detachment of three of 892 Squadron's Phantoms embarked for two weeks of trials. On 14 June 1970, the Phantom FG Mk 1s of No. 892 Squadron along with the Buccaneers S Mk 1s of No. 809, the Gannet AEW Mk 3s of No. 849 B Flight, and the supporting Sea King HAS Mk 1s and Wessex HAS Mk 1s – a total of 40 aircraft – embarked in HMS *Ark Royal* for six weeks in Home waters. In July 1972, 892 Squadron's shore base was transferred from RNAS Yeovilton to RAF Leuchars, which then became the home base of all the Phantom FG Mk 1s, including those belonging to the RAF. Over the nine years of HMS *Ark Royal*'s fifth commission, No. 892 Squadron was on board for some 29 cruises including trips to the Mediterranean, Caribbean and USA and during many of these there were cross-operations with the US Navy Carriers. HMS *Ark Royal*'s final voyage was to the USA during 1978 and the last aircraft to be catapulted from her was an 892 Squadron Phantom, XT870/012/R, on 27 November 1978. All the squadron's Phantoms were flown directly from *Ark Royal* to RAF St. Athan where they were handed over to the RAF and No. 892 Squadron formally disbanded on 15 December 1978.

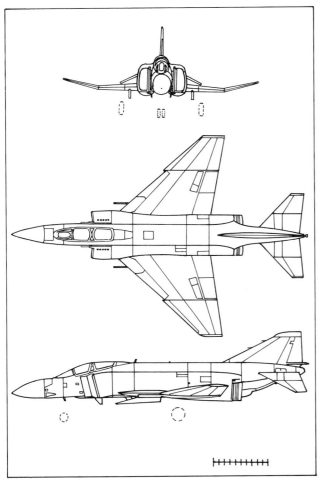

38   Phantom FG Mk 1

PRODUCTION

YF-4K   XT595, XT596.
(F-4K) FG Mk 1   XT597, XT598, XT857 – XT872, XV565 – XV570, XV586 – XV592.
NB.   All the other FG Mk 1s were delivered to the RAF, although several were subsequently loaned to the RN for No. 767 Squadron.

CHAPTER THIRTY-NINE

Westland/Sikorsky Sea King

In 1957 the US Navy had issued a requirement for an ASW helicopter with a long endurance and capable of combining the submarine 'hunter' and 'killer' roles. Up to that time helicopter manufacturers had had considerable difficulty in producing a helicopter that could operate in the dual role and to meet this requirement the Sikorsky Aircraft Division of the United Aircraft Corporation designed a new large helicopter. Designated the S61, this was to be powered by two General Electric T58 turbo-shaft engines and carry a weapons load of over 800lb in addition to all the latest submarine detection equipment and enough fuel for a minimum four hour sortie. Seven of these were ordered by the

Sea King HAS Mk 1, XV666/144/E, of No. 826 Squadron, HMS *Eagle,* landing on HMS *Albion* in September 1972.

US Navy in December 1957 for development purposes and the first aircraft, now identified as the YHSS-2, flew on 11 March 1959.

For the following two years an intensive programme of development and evaluation trials was carried out before production orders were placed by the US Navy for the HSS-2, later to be re-designated the SH-3A and named Sea King. Deliveries commenced in 1961 and subsequently over 200 were delivered to the US Navy. A much improved version of the Sea King, the SH-3D, had been developed by 1966, powered by two 1400shp General Electric T58-GE-10 engines operating through a new stronger gear box to handle the increased power. Improved sonar and doppler systems had also been introduced and to cater for the increased weight some strengthening of the structure had been carried out. In

Sea King HAS Mk 2, XV661/303/PW, of No. 819 Squadron, Prestwick, with Ailsa Craig in the background, February 1979.
(HMS *Gannet*)

all 74 of the SH-3Ds were supplied to the US Navy and proved to be a significant improvement over the earlier variants. During the early 1960s the S61 was also developed for the US Air Force as a long-range, medium-lift transport designated the CH-3C and this was further developed into the HH-3E, a combat-area rescue helicopter which, during service in Vietnam, was nicknamed "Jolly Green Giant", a name that subsequently was accepted officially.

The Westland Aircraft Company in Britain had been taking an interest in these developments from the outset and in 1959 had signed a licence to build the S61 although at the time, their plans were to produce it for commercial operators, as the Royal Navy had the Bristol Type 192 under development to cover their ASW requirement. However, in June 1966 when the Bristol Type 192 had been cancelled and the Westland WS58 Wessex had proved unable to fully meet the requirements, a contract was placed with Westlands for an anglicised version of the Sikorsky Sea King.

To speed up development of the British variant, Westland purchased four S61s from Sikorsky and these were transported to the UK by ship and assembled by Westland. The first was re-assembled on the docks at Avonmouth, registered G-ATYU and flown to Westlands at Yeovil in October 1966. The three remaining aircraft were supplied as kits of components and were assembled at Yeovil. G-ATYU, which soon received serial number XV370, was basically a standard American Sea King fitted with American equipment and powered by General Electric T58 engines while the other three, XV371 to XV373, had much of their American equipment replaced and were powered by de Havilland Gnome turbo-shaft engines. After completing its initial phase of trials, XV370 was also fitted with Gnome engines and on 8 September 1967 became the first Sea King to fly with these engines. As the other pre-production aircraft were completed they joined the development programme, although unfortunately the fourth aircraft, XV372, which was being used by Rolls Royce for engine development, was badly damaged in a very heavy landing following an engine flame-out. Although it was not repaired it was returned to

Westlands and used for experimental purposes. On completion of its development work XV370 was allotted to the Empire Test Pilots' School while XV371 and XV373 were retained for research work by the Royal Aircraft Establishment at Farnborough and Bedford respectively.

The first production Sea King HAS Mk 1, XV642, was flown for the first time on 7 May 1969 at Yeovil, and in July was used for deck landing trials aboard RFA *Engadine*. By July six production aircraft had been delivered to RNAS Culdrose and on 1 July 1969 No. 700S Flight, the Sea King Intensive Flying Trials Unit, was formed with Lt. Cdr. V. G. Sirett in command. By the time the unit disbanded on 29 May 1970 it had flown some 2700 hours with six aircraft, the majority of the trials being concentrated on the evaluation of the primary anti-submarine role.

The Westland Sea King HAS Mk 1 was basically the same as the Sikorsky SH-3D, retaining the watertight hull and with the main undercarriage retracting into sponsons mounted alongside the fuselage and the nose-wheel retracting into the fuselage. The Mk 1 also had a fixed tail wheel and, to improve stability in the water, inflatable buoyancy bags were fitted to the outside of each sponson. However, with the engines, gearbox and transmission being mounted in the top of the fuselage, the centre of gravity was very high. Consequently the Sea King could only remain afloat without capsizing in very calm conditions. Although the design of the fuselage did give the impression of a helicopter flying boat it was never intended to operate from water. The powerplants were two 1500shp Gnome H1400 turbo-shaft engines driving a Sikorsky five-bladed, all metal main rotor which incorporated automatic powered folding to facilitate storage aboard ships. To meet the requirement for a fully-integrated, all-weather, hunter-killer weapons system capable of operating independently of surface ships the HAS Mk 1 was equipped with a Plessey dunking sonar, a Marconi doppler navigation system, an Ecko search radar mounted in a dorsal radome and a Louis Newmark automatic flight control system. The crew comprised two pilots, a sonar operator and an observer. A variety of stores could be carried including marine markers, smoke floats, depth charges or homing torpedoes.

In January 1970 No. 706 Squadron, the Helicopter Advanced Flying Training Squadron, took delivery of

Sea King Mk 2 (AEW), XV714/385, of No. 849 Squadron. (Rolls Royce)

HMS *Invincible* in which it embarked a detachment of two aircraft on 18 November 1980, the remainder of the squadron following in February 1981. The Sea King Training Flight Element of No. 706 Squadron at RNAS Culdrose was the next unit to receive HAS Mk 5s which it operated alongside the remaining HAS Mk 2s. Subsequently all the squadrons that were operating HAS Mk 2s converted to HAS Mk 5s, the last one being No. 819 Squadron based at Prestwick which started to receive its Mk 5s in March 1985. A new Sea King squadron, No. 810, reformed at Culdrose on 15 February 1983 with ten Sea King HAS Mk 5s. The squadron took over from No. 737 Squadron the responsibility for advanced and operational flying training for ASW pilots, observers and aircrewmen.

Sea Kings made up a major part of the Fleet Air Arm's helicopter element of the Falklands Task Force, with a total of 50 aircraft being used by the five Sea King squadrons involved.

When HMS Invincible received the order to sail to the South Atlantic the nine Sea King HAS Mk 5s of

The first Sea King HAS Mk 6 ZA 136/251 to join the Sea King HAS Mk 6 Intensive Flying Trials Unit operated by No. 824 Squadron at HMS *Gannet*, Prestwick, flying in formation with one of the squadron's Sea King HAS Mk 5s in August 1988. (HMS *Gannet*)

No. 820 Squadron, under the command of Lt. Cdr. R. J. S. Wykes-Sneyd, had already embarked in readiness for the ship's next deployment. Before sailing the squadron strength was increased by two aircraft transferred from No. 706 Squadron as Front-line Immediate Replacement (FIR) aircraft. The long voyage south was used for a period of intensive training, including the use of live weapons and during this time, on 23 April 1982, a Sea King from No. 820 Squadron rescued the pilot of a Sea King HC Mk 4 of No. 846 which had ditched while flying from HMS *Hermes*, the other crew member being unfortunately lost. Because of the fear of Argentinian submarines rumoured to be patrolling the Total Exclusion Zone (TEZ), No. 820 Squadron's Sea Kings were providing regular ASW patrols and in addition, to protect the ship from air attack, the Sea Kings were also heavily involved in chaff-laying. On 2 June a Sea King, XZ574 flown by Lt. Cdr. K. Dudley, rescued Flt. Lt. Mortimer, RAF, from the sea after he had ejected from his stricken Sea Harrier.

No. 824 Squadron, commanded by Lt. Cdr. I. Thorpe and equipped with six Sea King HAS Mk 2s, was split up into three Flights, A, B and C, each of two aircraft, to operate from three RFA tankers. However, as only two of the tankers, RFA *Olmeda* with A Flight and RFA *Fort Grange* with B Flight sailed to the South Atlantic, only four of the squadron's Sea Kings took part in Operation Corporate.

To meet the demand for Sea King units, No. 825 Squadron was reformed, under the command of Lt. Cdr. H. S. Clark, specifically for Operation Corporate with ten Sea King HAS Mk 2s. Two of the Sea Kings embarked in RMS *Queen Elizabeth 2*, the remainder joining SS *Atlantic Causeway* and operating primarily in the trooping and heavy lift roles. A squadron detachment operated ashore at Port San Carlos in June but shortly after the end of hostilities the squadron re-embarked and returned to the UK in August to be disbanded at Culdrose on 17 September 1982.

Under the command of Lt. Cdr. D. J. S. Squier, No. 826 Squadron embarked its nine Sea King HAS Mk 5s aboard HMS *Hermes*. The squadron suffered its first loss on 12 May when ZA132 ditched following engine failure. Fortunately the crew were rescued after only a short time in the water. A second Sea King, XZ573, ditched on 18 May and remained upright and afloat.

Sea King Mk 2a (AEW), XV650/588/CU, the second development aircraft, still carrying the markings of No. 706 Squadron, its previous operator. (Westland)

The crew were picked up but as it was not considered wise to loiter with the risk of enemy submarines in the area attempts to salvage the aircraft were abandoned and it was sunk by naval gunfire. On 25 May the squadron provided Sea Kings to pick up survivors from the stricken HMS *Coventry* and the *Atlantic Conveyor*. Five days later XZ571 flown by S/Lt. K. B. Sutton rescued Sqn. Ldr. J. Pook, RAF, of No. 1(F) Squadron who had ejected from his Harrier GR Mk 3 after it had been hit by small arms fire. It is perhaps interesting to note that for a period of two months after HMS *Hermes* left Ascension Island on 18 April, the squadron kept at least three Sea Kings airborne at all times on ASW patrols. In January 1983 the squadron split into three flights of five aircraft each and for some years after the end of the Falklands War the squadron rotated the Flights so that there was always one in the South Atlantic.

The only first line squadron operating Sea King HC Mk 4s at the time of Operation Corporate, No. 846, embarked nine aircraft, under the command of Lt. Cdr. S. C. Thornewill, in HMS *Hermes,* while the remaining three aircraft under the command of the senior pilot embarked in HMS *Fearless.* A thirteenth HC Mk 4 was added to the squadron, this aircraft being delivered to Ascension in the hold of a Heavylift Belfast, where it was attached to the resident Naval Party 1222. The Sea King HC Mk 4 was the main supply and troop carrying helicopter in the Task Force and with the loss of virtually all the RAF's Chinooks on board *Atlantic Conveyor,* remained so after the landings. In addition to its routine activities, the squadron was also involved in various special duties, including transporting a unit of D Squadron, 22 SAS Regiment on a reconnaissance of Pebble Island during the night of 11 May following this up with a full scale raid on the night of 14 May when an ammunition dump and a number of aircraft were destroyed. Three of No. 846 Squadron's Sea Kings were lost during the Falklands campaign. The

first, ZA311, crashed into the sea on 23 April and although the pilot was rescued by a Sea King from No. 820 Squadron, the aircrewman was, unfortunately, lost.

In mid May it proved necessary to transfer men and equipment from HMS *Hermes* to HMS *Intrepid* to make space for an extra four No. 809 Squadron Sea Harrier FRS Mk 1s and four No. 1(F) RAF Squadron Harrier GR Mk 3s and three of No. 846's Sea Kings were used to carry out the transfer. While transferring 27 SAS troops from *Hermes* on 19 May, Sea King ZA294 suddenly lost power after having collided, it is believed, with a large bird, and crashed into the sea with the loss of 21 lives. At about the same time as this, two Sea King HC Mk 4s were transferred from HMS *Hermes* to HMS *Invincible* for a special long range reconnaissance mission. Subsequently one of these Sea Kings, ZA290, had to make a forced landing on a beach at Agua Fresea in Chile and was deliberately set on fire by the crew; presumably because of the clandestine nature of the mission. During the period of Operation Corporate, the Sea Kings of No. 846 Squadron had flown a total of over 3100 hours and had made over 3300 deck landings.

One of the most important results of the Falklands War was the highlighting of the Royal Navy's need for a replacement for the Gannet AEW Mk 3 in the Airborne Early Warning role. These aircraft had been taken out of service in 1978 with the withdrawal of HMS *Ark Royal*, the expectation being that the ships of the Royal Navy would be covered by the RAF's Shackleton AEW Mk 2s and then by their replacement, the Nimrod AEW Mk 3. The Falklands War showed that it was not always possible for land based aircraft to guarantee AEW cover for the fleet. To fill the gap Westland immediately set to work on converting two Sea King HAS Mk 2s, XV650 and XV704, to the AEW role. For this role Thorn-EMI Searchwater radar was installed in a large inflatable fabric radome mounted on the starboard side which could be rotated through 90 degrees to keep it clear of the ground for take-off and landing. With an urgency typical of the Falklands period, Westland achieved what under normal circumstances would have been considered impossible. Hav-

The first prototype EH101, ZF641, flying over Yeovil. (Westland)

ing received the two aircraft at the end of May both were converted and flying by the end of July. These aircraft, designated the Sea King HAS Mk 2 (AEW), were flown aboard the South Atlantic bound HMS *Illustrious* on 2 August, becoming the equipment of D Flight of No. 824 Squadron; which had been formed specifically for the AEW Role on 14 June 1982 at RNAS Culdrose. Although only pre-production, development aircraft, and not to full production standard, they proved to be very satisfactory and orders were placed with Westlands and the Royal Naval Aircraft Yard at Fleetlands to convert a further six Sea King HAS Mk 2s into AEW Mk 2s and also to bring the two development aircraft up to full production standard. On 1 November 1984, No. 849 Squadron reformed at RNAS Culdrose from No. 824D Flight and equipped with the two pre-production aircraft. A Flight, the first operational flight, was commissioned on 31 May 1985 with the first three production Sea King AEW Mk 2s and embarked in *Illustrious* during August. As it is expected that only two of the ASW carriers will be operational at any one time, it is planned to have only two operational flights and a headquarters flight.

The latest version of the Sea King is the HAS Mk 6 which introduces improvements to the airframe and main transmission, in addition to major modifications to the ASW sonics, sonar and MAD systems. A contract for four new build HAS Mk 6s was placed with Westland Helicopters in 1987 and this was followed shortly afterwards by a contract to convert 25 of the Royal Navy's HAS Mk 5s into Mk 6s. During 1988 No. 824 Squadron at HMS *Gannet,* Prestwick is to form the Sea King HAS Mk 6 Intensive Flying Trials Unit.

The 7 April 1987 saw the roll-out at Westland's Yeovil factory of the EH101, the intended successor to the Sea King. Two days later the British Government announced that orders had been placed for 50 of the ASW version of the EH101 for the Royal Navy, plus 25 of a utility version for use primarily as troop transports. The EH101 project started in 1979 following discussions between Westland and Agusta of Italy about the joint development and production of a replacement for the Sea King in both the military and commercial roles. The British and Italian governments signed a Memorandum of Understanding in November 1979 and, in June 1980, EH Industries was formed by the two companies to control the project. A second Memorandum of Understanding was signed in June 1981, at the beginning of the government funded project definition stage. The full development stage commenced in 1983 and culminated in the roll-out and first flight of the first pre-production prototype EH101, ZF641, on 9 October 1987.

The EH101 is designed for fully autonomous all-weather day and night operations from land bases and ships, including being capable of launch from and recovery to a ship as small as a frigate in wind speeds of up to 50 knots. Power is to be provided by three General Electric T700-401 turbine engines driving the latest, state of the art, composite, five blade main rotor. It is however expected that later production aircraft will be powered by the Rolls-Royce Turbomeca RTM 322 engine. The Royal Navy's ASW/ASV variant of the EH101 will be capable of carrying out all the roles currently being undertaken by the Sea King and will be equipped with the latest avionic kit available including the Ferranti Blue Kestrel search radar mounted in a chin radome, dipping sonar and sonobuoys, Racal ESM and up to four homing torpedoes and anti-shipping missiles. The EH101 looks as though it will be a worthy successor to the excellent Sea King but as it is unlikely to enter service before 1992, the Sea King will almost certainly remain in service with the Fleet Air Arm into the 21st century.

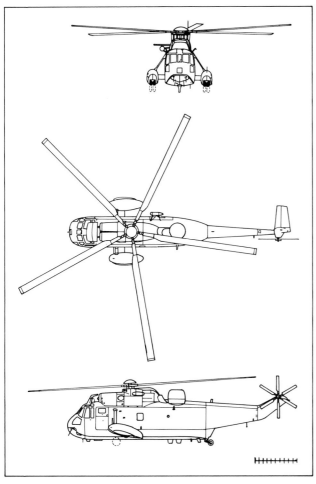

39 Sea King HAS Mk 5

PRODUCTION

HAS Mk 1 (Prototypes) XV370 – XV372 – Supplied by Sikorsky.

HAS Mk 1/2A XV642 – XV677, XV695 – XV714.

HAS Mk 2 XZ570 – XZ582, XZ915 – XZ922.

HAS Mk 5 ZA126 – ZA137, ZA166 – ZA170, ZD630 – ZD637, ZE418 – ZE422.

MC Mk 4 ZA290 – ZA299, ZA310 – ZA314, ZD476 – ZD480, ZD625 – ZD627, ZE425 – ZE428, ZF116 – ZF124.

AEW Mk 2 (Conversions) XV650, XV704 (development), XV649, XV656, XV671, XV672, XV697, XV714.

CHAPTER FORTY
Aerospatiale/Westland Gazelle

The Gazelle was the second of the helicopters covered by the Anglo/French collaborative agreement that had been signed in February 1967. The first two types, the Puma and the Gazelle, were both of French origin with only the third, the Lynx, being British. Design of the Gazelle had started at Aerospatiale in the early 1960s, basically as an Alouette replacement, to meet a French Army requirement for a light observation helicopter. The initial project, identified as the X300, was a small light observation helicopter but this was considered by Aerospatiale to be rather too specialised and to make it more versatile it was necessary to increase the size.

Consequently the next stage of development, the SA340, introduced a much larger cabin capable of carrying three passengers or two stretcher cases, in addition to the crew of two. Although consideration was given to various powerplants including the Continental T65, Allison T63 and AiResearch TSE331, Aerospatiale finally decided to use the Turbomeca Astazou turboshaft engine. To speed development the first SA340 Gazelle prototype was fitted with the same Astazou engine and transmission as the Alouette 2, and this aircraft flew for the first time at Marignane in France on 7 April 1967.

The second prototype which was completed in 1968 showed some quite radical changes from the first, with the introduction of a tall fin in the base of which was

Three Gazelle HT Mk 2s, including XW861/559/CU, of No. 705 Squadron's 'Sharks' display team. (RNAS Culdrose)

Gazelle HT Mk 2, XW891/549/CU, of No. 705 Squadron's 'Sharks' display team. (RNAS Culdrose)

mounted a new ducted, thirteen-bladed tail rotor. An Astazou 11N2 engine driving a new type composite main rotor was also introduced. Four pre-production aircraft, identified as SA341s, were then built in France, with the first flying on 2 August 1968, and although the second and fourth of the pre-production batch were retained by Aerospatiale in France for development flying, the third was delivered to Westland's. There it was used as the prototype Gazelle AH Mk 1. It flew for the first time on 28 April 1970 bearing the military serial number XW276. On 14 May 1971 the first pre-production SA341 was used to set up three new world speed records for the E1c class of helicopter, achieving 192.6 mph over 3km, 193.9 mph over a 15/25km course and 183.9 mph for the 100km closed circuit.

Production of the Gazelle was soon under way in both France and Britain with the first flying at Marignane in France on 6 August 1971 followed, on 31 January 1972, by the first Westland built Gazelle. Although assembly lines were set up at both Aerospatiale and Westland, the agreement split the manufacture of components between the two manufacturers on a man hour basis, in proportion to the quantity of Gazelles purchased by each country and with a minimum of duplication of manufacturing effort.

The Gazelle AH Mk 1 was ordered in quantity for the British Army and the Royal Marines with 203 being delivered to the Army and nine to the Royal Marines. Although designed specifically for Army use, both the RAF and the Royal Navy considered that it would be ideal as a trainer and, at an early stage of production, orders were placed with Westlands for 35 HT Mk 2s for the Royal Navy and 33 HT Mk 3s for the RAF, as well as one for the Empire Test Pilot's School. At a later date an order was placed for a single communications version, the HCC Mk 4, which was to be operated by No. 32 Squadron RAF at Northolt.

The Fleet Air Arm's Gazelle HT Mk 2 saw the introduction of a stability augmentation system and a rescue hoist, both fairly basic requirements on Fleet Air Arm helicopters and consequently essential for its training role. The first HT Mk 2, XW845, flew for the first time in July 1972. The type entered service with No. 705 Squadron, commanded by Lt. Cdr. C. J. S. Craig, at RNAS Culdrose during December 1974, replacing its existing Whirlwind HAS Mk 7s and Hiller HT Mk 2s. This squadron has an excellent formation display team, called the "Sharks", comprising six Gazelle HT Mk 2s which performs at air displays throughout the British Isles during the summer months. The Gazelle has given excellent service over the past decade, and there appear to be no plans to replace it in the foreseeable future.

PRODUCTION

HT Mk 2 XW845, XW853, XW854, XW856, XW857, XW859 – XW861, XW863, XW864, XW867, XW868, XW871, XW884, XW886, XW887, XW890, XW891, XW894, XW895, XW907, XX391, XX397, XX410, XX415, XX431, XX436, XX441, XX446, XX451, XZ938, XZ942, ZB647, ZB648, ZB649.

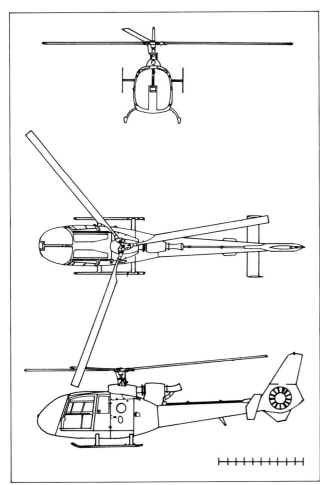

40   Gazelle HT Mk 2

Lynx HAS Mk 2, XZ238/646, of No. 702 Squadron, Portland, carrying Sea Skua air-to-surface missiles. (HMS *Osprey*)

was XX153, the first prototype of the Army utility Lynx AH Mk 1, which flew for the first time on 12 April 1972. Shortly afterwards on 20 June XX153, flown by Roy Moxam, set a new world speed record in the E1e Class of 199.92 mph. On 22 June the same pilot and aircraft achieved 197.914 mph over a 100km closed circuit course.

The first Naval development Lynx HAS Mk 2, XX469, flew for the first time on 25 May 1972 and, in addition to being used for weapons and equipment testing, was also used for deck landing trials on board HMS *Engadine* before being destroyed in a landing accident, caused by the loss of tail rotor control, in November 1972. Development flying with the remainder of the development aircraft continued without further incident and in May 1974 the first production contract was placed for 60 Lynx HAS Mk 2s for the Royal Navy together with quantities of Lynx AH Mk 1s for the British Army, MK 2(FN)s for the French Navy and HAS Mk 25s for the Royal Netherlands Navy.

The first production aircraft was a Lynx HAS Mk 2, XZ227, for the Royal Navy which flew for the first time on 10 February 1976. Deliveries to the Royal Navy commenced in August 1976 and on 1 September 1976 the Lynx Intensive Trials Unit (IFTU), No. 700L Squadron, commanded by Lt. Cdr. G. A. Cavalier, was formed as a joint Royal Navy/Royal Netherlands Navy Unit at RNAS Yeovilton. The unit initially operated seven Lynx but this was subsequently increased to eight. The first deck landing by a production Lynx was carried out by Lt. Cdr. Cavalier when he landed one of the unit's aircraft, XZ231, on HMS *Birmingham* during February 1977. The sea trials were subsequently completed using the ill fated HMS *Sheffield,* which was to be sunk some five years later by an Exocet missile during Operation Corporate in the Falklands. On completion of the trials 700L Squadron was disbanded on 16 December 1977 and formed the nucleus of No. 702 Squadron, commanded by Lt. Cdr. B. F. Prendergast and reformed at RNAS Yeovilton on 3 January

1978. No. 702 Squadron was the Lynx headquarters and training squadron being responsible for Lynx pilot and observer conversion and for advanced and operational flying training. In addition the squadron acted as headquarters for the 15 Lynx ship's flights that were now operating from destroyers and frigates; although the older Type 12 Rothesay-class frigates were unable to replace their Westland Wasps, their hangars being too small for the Lynx.

In the meantime Rolls Royce had continued development of the BS360 Gem engine beyond the 900shp Gem 2 used in the production Lynx. The development version of this uprated engine, identified as the Gem 4, was achieving 1050shp on the test bed and the production version, known as the Gem 41, with a little further development had increased its rating to 1120shp. Following the availability of the new engine the Royal Navy ordered a further batch of 20 Lynx powered by the Gem 41 engine and identified as the HAS Mk 3. To cope with the increased power a new strengthened transmission system was introduced, and the opportunity was taken to update the systems, the main change being the introduction of a passive Racal-Decca MIR-2 (Orange Crop) electronic surveillance system, also retrospectively fitted into the Lynx HAS Mk 2s. The first Lynx HAS Mk 3, ZD249, made its maiden flight on 26 August 1980 and deliveries to the Royal Navy commenced in March 1982.

The Royal Navy Lynx were equipped with the Ferranti Seaspray Mk 1 search and tracking radar, with the scanner located in the nose, for the detection of small high speed vessels. For naval operations a variety of weapons could be carried, including two Mk 44 or Mk 46 torpedoes, two Mk 11 depth charges or four Sea Skua anti-ship missiles.

On 1 January 1981 No. 815 Squadron, commanded by Lt. Cdr. D. H. N. Yates, was formed to take over from No. 702 the role of headquarters for the ship's flights. No. 702 retained its training role and continued to operate alongside No. 815 initially at Yeovilton and then at Portland to which both squadrons were transferred on 19 July 1982.

The Lynx HAS Mk 3 had not started to enter

operational service when the Falkland Islands and South Georgia were invaded by Argentinian forces on 2 April 1982, so the Lynx HAS Mk 2 was the standard helicopter for the ship's flights on the majority of destroyers and frigates that sailed south as part of the Royal Navy Task Force. In addition No. 815 Squadron also provided Lynx flights for the two carriers HMS *Hermes* and HMS *Invincible*. Although at this time the British Aerospace Sea Skua missile had not completed its acceptance trials, it was issued to the Task Force for use on the Lynx.

It was the Lynx from HMS *Brilliant* that was the first to see action during the retaking of South Georgia, by following up attacks on the Argentinian submarine *Santa Fe* by a Wasp and Wessex. The attack with a Mk 46 torpedo failed but some minor damage was achieved by machine gun fire. The *Santa Fe* was finally forced to beach after being damaged by Aerospatiale AS12 missiles launched by two Wasps. On 3 May a Sea King came under fire, and a Lynx armed with two Sea Skua missiles was sent from HMS *Coventry* to assist. Having located the 800 ton Argentinian ship *Somellera,* the Lynx fired its missiles from a range of some eight miles and the ship sank shortly after being hit by both missiles. A Lynx despatched from HMS *Glasgow* to search for survivors came under fire from a second Argentine ship, the *Alferez Sobral.* After moving out of range of the guns the Lynx launched both its Sea Skua missiles from a range of about nine miles and although the ship was not sunk it suffered severe damage when it was hit by both missiles. Another ship, the freighter *Rio Carcarana,* after being disabled by Sea Harriers was finally sunk by Sea Skua missiles, this time launched by the Lynx of HMS *Antelope*'s ship's flight.

During the Falklands War the Lynx amassed some 3050 flying hours during 1,863 sorties for the loss of three aircraft, which were all on board ships that were sunk during air attacks, XZ252 on HMS *Ardent*, XZ270 on *Atlantic Conveyor* and XZ242 on HMS *Coventry*.

An additional three Lynx HAS Mk 3s were ordered to replace the ones lost in the South Atlantic, and in mid 1985 an order was placed for a further seven, bringing the total quantity of production Lynx for the Royal Navy to 90.

Lynx HAS Mk 3, ZD252/302/PO, of No. 815 Squadron, Portland, carrying Sea Skua air-to-surface missiles. (HMS *Osprey*)

The Lynx has proved to be quite a success for Westland/Aerospatiale with 343 ordered to date. This includes, in addition to the 90 ordered for the Royal Navy, 128 for the British Army and Royal Marines and 112 for export to nine different countries. Of these export Lynx, 109 were basically naval versions similar to the HAS Mk 2 or Mk 3, while the remaining three, ordered for the Qatar police, were similar to the Lynx AH Mk 1 utility helicopter as used by the British Army.

In 1982 Westland started a programme, as a private venture, to revise and update the Lynx by introducing the latest technology. Although consideration had been given to a naval version, identified by Westland as the Super Lynx, to the extent of producing a full scale mock-up, Westland's efforts initially concentrated on a battlefield assault helicopter that will be available in the late 1980s. This range of helicopters called the Lynx 3 is powered by two 1260shp Rolls Royce Gem 60-3/1 engines; the airframe has been modified to introduce a longer nose and much thicker tailboom; pilot's vision has been improved; fuel capacity has been increased by some 35% and hard points have been introduced to cater for a wide variety of weapons. To speed development a Lynx AH Mk 1, ZE477, was converted on the production line to act as the Lynx 3 prototype. Flown for the first time on 14 June 1984, ZE477 was initially fitted with standard metal rotor blades but these are to be replaced by the new composite blades developed by Westland under the British Experimental Rotor Programme (BERP). The use of carbon fibre composites has enabled Westland to manufacture rotor blades of optimum profile design which includes the distinctive swept tip. These new rotor blades have provided an increased fatigue life and a significantly improved performance.

Development of the Super Lynx was continued as a private venture by Westland Helicopters, and ZD249 the first HAS Mk 3 was retained for development purposes. Early in 1987 the Royal Navy began to take an increasing interest in the Super Lynx and towards the end of 1987 a version of the Super Lynx, identified as the Lynx HAS Mk 8 was being developed for the Royal Navy. The current plan is to convert a number of the Royal Navy's existing Lynx helicopters to HAS Mk 8 standard by the introduction of Rolls Royce Gem 42-2 engines, Sea Owl passive identification device (PID) on the nose, Sea Spray Mk 3 360 degree radar scanner

Lynx HAS Mk 3, ZD249, being used by Westland for Super Lynx development. Note the HAS Mk 8 type 360 degree radar radome under the nose. (Westland)

mounted in a cylindrical radome under the nose and a secure speech system. Improvements will also be made to various other systems in addition to some structural strengthening.

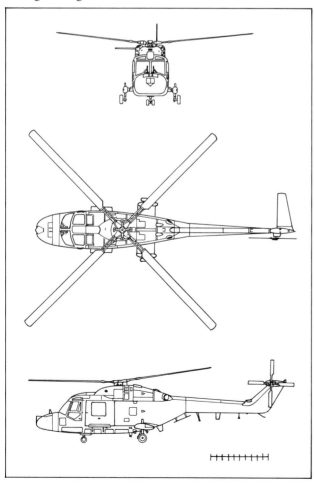

41 Lynx HAS Mk 3

With the availability of the new uprated Gem engine and the new rotor, Westland started to make plans to regain the world speed record for helicopters which had been held by Lynx XX153 in June 1972 but was currently held by the Russian MIL A10 Hind. To mount its challenge for the record the company selected the Lynx demonstrator, G-LYNX, which had been in storage for two years. This was modified by the introduction of the Gem 60 engine, fitted with water-methanol injection and driving a rotor which comprised a standard hub fitted with BERP 111 composite main rotor blades. To maintain stability the areas of both the horizontal and vertical stabilisers were increased. Flown by Trevor Egginton, Westland's chief test pilot, on a 15km course over Sedgemoor on 11 August 1986 G-LYNX set a new world record of 249.10 mph, breaking the existing record by 21 mph.

The Royal Navy's Lynx has proved invaluable, particularly in the ASV role and are expected to remain in service until at least the end of the century.

PRODUCTION

Prototypes   XX469, XX510, XX910, XZ166.
HAS Mk 2   XZ227 – XZ252, XZ254 – XZ257, XZ689 – XZ700, XZ719 – XZ736.
HAS Mk 3   ZD249 – ZD268, ZD565 – ZD567.

CHAPTER FORTY-TWO

Handley Page Jetstream

An apparent market, particularly in the USA, for a small modern feeder-liner was identified by Handley Page Ltd. in the early 1960s. A series of project studies resulted in the proposed HP137 a streamlined, pressurised aircraft capable of carrying up to 18 passengers and to be powered by two turbo-prop engines. The manufacture of four prototypes was authorised in January 1966 and the first, G-ATXH, powered by 690shp Turbomeca Astazou XII engines, made its maiden flight at Radlett on 18 August 1967 with Handley Page's chief test pilot, John Allam, at the controls. At an early stage, dealerships were set up in the USA and UK and by the time of the first flight orders had been received from them for 165 of the aircraft, by then named the Jetstream.

During 1968 Handley Page were successful in winning an order for eleven Jetstreams for the USAF. These aircraft, designated the C-10A by the USAF, were to introduce a strengthened wing and Garrett TPE331 engines and an early production Jetstream, converted to act as a prototype of the C-10A, was identified as the Jetstream 3M.

Handley Page started to run into serious cash flow problems in 1969 and although several attempts were made to save the company, it was declared bankrupt in 1970 and all work on the Jetstream was stopped after

70 aircraft, including the prototypes, had flown. To provide support for aircraft already in service a new company, Jetstream Aircraft Ltd., was formed the following year and in 1972 when the RAF was showing an interest in the Jetstream for multi-engined training, this company, together with Scottish Aviation, formed Scottish Jetstream Ltd. to handle the project. When an order for 26 Jetstreams for the RAF was eventually placed, Scottish Aviation took over Jetstream Aircraft Ltd. and with it the full responsibility for the Jetstream.

Of the 26 Jetstreams built for the RAF, and known as the T Mk 1, only five were in effect totally new build aircraft as two aircraft were purchased in the USA and converted and 19 were built from partly completed airframes and components bought from the Handley Page receiver by Terravia Trading Service Ltd. and stored by them until acquired by Scottish Aviation. The first Jetstream T Mk 1 for the RAF flew for the first time at Prestwick on 13 April 1973. The Jetstreams were operated by the Central Flying School at Little Rissington and No. 5 Flying Training School, but in 1975 they were withdrawn from use and put into store. Coincidentally, at this time the Royal Navy was starting to look for a replacement for the elderly Sea Princes of No. 750 Squadron, used primarily for observer training. Initially, a single Jetstream T Mk 1, XX475 on loan from the RAF, was used for evaluation by No. 750 Squadon during 1976. However it was not until the end

Jetstream T Mk 2, ZA111/574/CU, of No. 750 Squadron, Culdrose. (RNAS Culdrose)

Jetstream T Mk 3, ZE438/576, of No. 750 Squadron, Culdrose. (RNAS Culdrose)

of 1977 that it was decided to develop the Jetstream for the observer training role.

The Royal Navy had a requirement for 16 Jetstreams to replace its Sea Princes, and these aircraft were to be conversions of the RAF's Jetstream T Mk 1s still in store. Unfortunately at the same time the RAF decided to restart their multi-engined training programme, requiring eleven of the stored Jetstreams and, as one had been written-off during their previous period of service, only 14 were available to meet the Royal Navy's requirement. To make up the shortfall two civil aircraft had to be obtained for conversion. The Royal Navy's version of the Jetstream, identified as the T Mk 2, differed from the T Mk 1 primarily by the equipment fit which included a Tactical Air Navigation System (TANS), Decca Doppler 71, and MEL E190 terrain and weather mapping radar. The 18in radar scanner was housed in a conspicuous nose mounted thimble radome and the doppler aerial was accommodated in a fairing under the fuselage centre section. For the students, two fully instrumented navigation consoles were installed in the cabin. The first Jetstream T Mk 2, XX481, was delivered to No. 750 Squadron at RNAS Culdrose on 26 October 1978 and by May of the following year all the Sea Princes had been replaced.

Early in 1984 an additional four Jetstreams were ordered for the Royal Navy, and these were basically Jetstream 31s, the latest version, powered by Garrett TPE331 turboprop engines. This version of the Jetstream, designated the T Mk 3, was fitted with the Racal ASR360 radar with its ventrally mounted radar scanner displacing the nose mounted MEL radar and consequently changing the distinctive nose shape of the T Mk 2. The Jetstream T Mk 3s were completed towards the end of 1985 and flew initially carrying the company's class 'B' markings before their military serial numbers ZE438 to ZE441 were applied to the airframes. Deliveries to No. 750 Squadron at Culdrose started early in 1986, where they augmented the existing Jetstream T Mk 2 fleet.

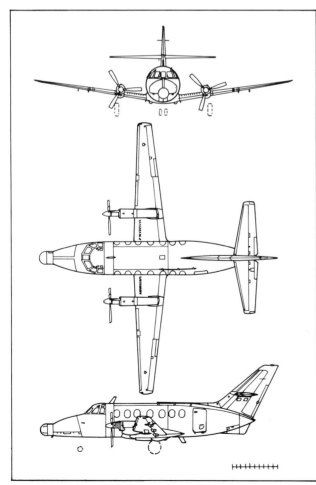

42 Jetstream T Mk 2

PRODUCTION

T Mk 2   XX475, XX476, XX478 – XX481, XX483 – XX490, ZA110, ZA111.
T Mk 3   ZE438 – ZE441.

CHAPTER FORTY-THREE
British Aerospace Sea Harrier

The Hawker Aircraft Company started to take an interest in vertical take-off and landing (VTOL) aircraft in 1957 with two projects, the P1126 which was to use twelve lift engines for VTOL operation, and the much more radical P1127 which was to be powered by a single vectored-thrust engine. Hawker's envisaged the use of the Bristol BE53 engine but as the early scheme was to have only two rotating nozzles they believed that they would have serious problems with stability in the landing and take-off modes. The Hawker design team came forward with a scheme that would introduce an additional pair of rotating nozzles and following this, Bristol revised the design of the BE53 which was re-designated the BE53/2. Manufacture of prototype

Sea Harrier FRS Mk 1, XZ750, during type operational trials on HMS *Hermes* in October 1979. (B.Ae.)

engines, which by then had been named Pegasus 1, was put in hand and the first engine ran for the first time in September 1959.

Before this, in March 1959, Hawker Aircraft had decided to proceed with the manufacture of two experimental prototypes of the P1127 as a private venture because of the lack of official interest in either the engine or airframe at that time. However some time later the RAF did start to show some interest in the project, as a possible replacement for the Hunter in the tactical ground support role. The Ministry of Aviation finally took over the project in June 1960 and provided a contract for the first two prototypes which were then being built. These two aircraft were allocated serial numbers XP831 and XP836.

On 21 October 1960 XP831, with A. W. Bedford Hawker's chief test pilot at the controls, hovered for

Harrier T Mk 4N, ZB605/718, of No. 899 Squadron, Yeovilton, 1985. (B.Ae.)

the fifth production Harrier GR Mk 1 operated from the helicopter platform of the cruiser HMS *Blake*. Service release for Harrier deck operations by RAF pilots was issued in March 1970 following trials aboard HMS *Eagle* by two Harrier Mk 1s and in 1971 No. 1(F) Squadron embarked in HMS *Ark Royal* for a cruise in Home waters.

Following the loss of the CVA-01, the Admiralty started planning a new type of ship, the anti-submarine cruiser, sometimes referred to as the through deck cruiser. The final design of this vessel can best be described as a small aircraft carrier, and at the time it was only intended for helicopter operations. Three of these anti-submarine cruisers were ordered in the early 1970s and all three, *Invincible, Illustrious* and *Ark Royal,* entered service ten years later, by which time they were called ASW carriers.

The Admiralty had started in 1971 to prepare a requirement for a maritime Harrier to operate from the commando carriers or the planned ASW carriers.

The Fleet Air Arm required a much more versatile Harrier than had been built at the time, capable of operating in the air defence as well as the reconnaissance, strike and ground attack roles. To meet this requirement there were many changes to the Harrier's avionic fit including the introduction of a new Smiths Industries head up display along with the Ferranti Blue Fox radar. With the changes to the nose necessary to accommodate the Blue Fox radar and the redesign of the cockpit to take the extra naval equipment the opportunity was taken to raise the cockpit, providing much more space for equipment and considerably improving the pilot's forward view. A contract was placed in May 1975 for three development and 31 production aircraft which by then had been named the Sea Harrier FRS Mk 1. To assist with the development programme three Hunter T Mk 8s were given the full Harrier avionics fit including the Blue Fox radar, and on completion of the development programme these aircraft were handed over to the Fleet Air Arm for use in the training role and were re-designated the Hunter T Mk 8M.

The first Sea Harrier FRS Mk 1 to fly was the first production aircraft, XZ450 which, because of its limited instrumentation fit, was completed before the development aircraft and was demonstrated at the 1978 SBAC Display by John Farley. By June 1979 the three development aircraft, XZ438, XZ439 and XZ440, had flown, as had the second production aircraft XZ451. Subsequently an order was placed for three Harrier T

Mk 4N two-seat trainers to assist with pilot conversion and the first of these, ZB604, was delivered to Yeovilton on 21 September 1983.

On 26 June 1979, No. 700A Flight, the Sea Harrier Intensive Flying Trials Unit, had formed with a single Sea Harrier, XZ451, at RNAS Yeovilton under the command of Lt. Cdr. N. D. Ward. By January 1980 the flight was operating its full complement of six aircraft, and had completed the type operational trials on HMS *Hermes*. When the flight disbanded on 31 March 1980 it immediately reformed as No. 899 Squadron, the Sea Harrier Headquarters Squadron, still under the command of Lt. Cdr. Ward. This squadron in its secondary training function also operated a two-seat Harrier T Mk 4N and one of the Hunter T Mk 8Ms. No. 800 Squadron, commanded by Lt. Cdr. T. J. H. Gedge, re-formed at RNAS Yeovilton on 31 March 1980 with four Sea Harrier FRS Mk 1s and served in HMS *Invincible* until June 1981 when it transferred to HMS *Hermes* and later, in September 1983, moved to HMS *Illustrious*.

There have been only two other first line Sea Harrier squadrons, No. 801 which, under the command of Lt. Cdr. N. D. Ward, had reformed at RNAS Yeovilton on 28 January 1981 for service in HMS *Invincible,* and finally No. 809 Squadron under the command of Lt. Cdr. T. J. H. Gedge which had reformed on 8 April 1982 out of No. 899 Squadron specifically for service with the Falklands Task Force. Although No. 809 was expected to disband at the end of hostilities in the Falklands, after returning home it joined HMS *Illustrious* on her first operational voyage, which was ironically to the South Atlantic. The squadron disbanded on 17 December 1982 after *Illustrious*'s return to the UK.

When the Task Force sailed from the UK on 5 April 1982 to recover the Falkland Islands from the Argentine invaders, it was led by HMS *Hermes*, which had been refitted for the dual ASW/Commando role during 1976/77. For Operation Corporate the complements of the Sea Harrier Squadrons had been increased, which had stretched the Fleet Air Arm's meagre resources of aircraft and pilots to the limit. However, No. 800 Squadron, commanded by Lt. Cdr. A. D. Auld, which had been given a complement of twelve aircraft, had been allotted to HMS *Hermes* and had managed to embark eleven aircraft before the ship sailed. The last aircraft was flown aboard by an RAF pilot as *Hermes* sailed down the English Channel. The only ASW carrier available at the start of the Falklands War was HMS *Invincible,* which being rather smaller than *Hermes* embarked the eight Sea Harriers of No. 801 Squadron under the command of Lt. Cdr. N. D. Ward.

The third squadron to take part in Operation Corporate was No. 809 which was formed to provide a back up of aircraft and pilots for both No. 800 and No. 801 Squadrons. Insufficient Sea Harriers were available to achieve the squadron's full complement of ten aircraft. Six of No. 809 Squadron's Sea Harriers left Yeovilton on 30 April, the remaining two leaving the next day for the flight to Wideawake on the Ascension Islands where all eight embarked on the ill-fated RO-RO Container ship, *Atlantic Conveyor*. On 8 May, some two weeks before the *Atlantic Conveyor* was lost in an Exocet attack, five of these aircraft were flown to *Hermes* with one flying on to *Invincible*, and the remaining three also flew to *Invincible* the following day.

Initially to provide back up for the Fleet Air Arm Sea Harriers, No. 1 Squadron RAF, equipped with 14 Harrier GR Mk 3s, was also sent to the South Atlantic and, like No. 809 Squadron, first flew direct to Ascension Island to embark in *Atlantic Conveyor* for the voyage to the Falklands. There the squadron transferred to HMS *Hermes* which was to be its base for operations against Argentinian ground targets on the Islands.

The Sea Harriers first went into action during the morning of 1 May, when HMS *Hermes* launched the twelve Sea Harriers of No. 800 Squadron, nine to attack the airfield at Port Stanley and three the airstrip at Goose Green. Top cover for the attacks was provided by Sea Harriers of No. 801 Squadron launched on Combat Air Patrols (CAP) from HMS *Invincible*. Although encountering defensive fire of high intensity, the attacks on the airfields with rockets and 1000lb bombs were very successful with several parked aircraft and much of the airfield equipment being damaged or destroyed, the only damage suffered by the attacking

Sea Harriers being a small hole in a fin. During that first day the Sea Harriers of No. 801 Squadron on CAP had made their first contact with enemy aircraft and during the first combat Flt. Lt. P. Barton, RAF, flying XZ452, had destroyed a Mirage 111EA with an AIM-9L Sidewinder air-to-air missile, and during the same combat a second Mirage had been damaged by a Sidewinder from XZ453 flown by Lt. S. Thomas. This crippled aircraft suffered the indignity of being shot down by its own side's anti-aircraft fire while trying to make an emergency landing at Port Stanley.

Later in the day No. 801 Squadron claimed its third kill when Lt. A. Curtis shot down a low flying Canberra and No. 800 Squadron gained its first when a Dagger – an Israeli built Mirage 5 – was destroyed by a Sidewinder launched from Sea Harrier XZ455 flown by Flt. Lt. R. Penfold, RAF. On the following day the first Sea Harrier was lost when XZ450 of No. 800 Squadron was hit by anti-aircraft fire and exploded during an attack on the Goose Green airstrip, unfortunately killing the pilot.

On 6 May two more Sea Harriers, XZ452 and XZ453 of No. 801 Squadron, were lost when they flew into low cloud and fog while investigating a radar contact, disappeared without trace and were presumed to have collided. By the time of the Argentinian surrender on 14 June a further three Sea Harriers had been lost, but only one of them in combat. ZA192 of No. 800 Squadron crashed into the sea on 23 May shortly after a night take-off from HMS *Hermes* and ZA174 of No. 801 Squadron, while positioning for take-off on the deck of HMS *Invincible* during a storm, slid into the sea off the rolling ship. In this case the pilot ejected and was picked up by the SAR helicopter. The final Sea Harrier to be lost was XZ456 of No. 801 Squadron which was hit by a Roland surface-to-air missile while on a reconnaissance mission in the vicinity of Port Stanley Airport, the pilot ejected and was picked up from the sea by a Sea King from HMS *Invincible*.

Line-up of Sea Harrier FRS Mk 1s of No. 809 Squadron, Yeovilton, commissioned to meet the operational demands of Operation Corporate. (B.Ae.)

Sea Harrier FRS Mk 1, XZ495/003/N, of No. 801 Squadron coming in to land on HMS *Invincible*. (Westland)

It is interesting to note that of the six Sea Harriers lost only two were the result of enemy action, both to ground fire, the other four being the result of accidents. This compares with a total of 24 Argentinian aircraft shot down by Sea Harriers, the majority of those being combat aircraft, Daggers (Mirage 5), Mirage 111EAs and Skyhawks. In addition to the aircraft destroyed in the air quite a number were destroyed on the ground by the Sea Harriers and the RAF's Harrier GR Mk 3s during their attacks on the airfields, which also caused major disruptions to the Argentinian air operations. The Sea Harriers were also involved in attacks on Argentinian shipping, starting with the stern trawler *Narwal* which on 9 May was located some 60 miles from the Falklands and was attacked by four of No. 800 Squadron's Sea Harriers, whose bombing and strafing attacks put her out of action. She was subsequently captured by the Royal Navy but sank while under tow. A further two ships, the *Bahia Buen Suceso* and *Rio Carcarana*, were attacked by No. 800 Squadron on 16 May and both ships were put out of action.

The Falklands War provided the opportunity for the Sea Harriers to demonstrate their excellence and during the short period of the hostilities the aircraft had flown 2376 sorties totalling some 2675 flying hours and there is no doubt that without them there would have been little prospect of Britain regaining the Falkland Islands.

With the ending of hostilities the Task Force progressively returned to the UK, leaving behind a military presence large enough to deter any future invader. No. 809 Squadron, which had been used during the conflict to provide aircraft and pilots to supplement Nos. 800

and 801 Squadrons, was again re-established in its own right as an operational squadron with ten Sea Harrier FRS Mk 1s and embarked in the new HMS *Illustrious* for her maiden operational patrol in the South Atlantic from August until December 1982. No. 809 Squadron disbanded at Yeovilton shortly after returning to the UK. During 1982 orders were placed for a further 14 Sea Harriers to increase the basic strengths of the squadrons as well as replace the aircraft lost in the Falklands. A further nine were ordered in September 1984 bringing the total to 57. Rather surprisingly attempts to sell the Sea Harrier abroad have met with only limited success, six Sea Harrier FRS Mk 51s being sold to the Indian Navy to replace Sea Hawks in December 1980. There are also hopes that Italy will order a batch to operate from its helicopter carrier *Guiseppe Garibaldi*.

During the Falklands campaign, a shortcoming of the Blue Fox radar was encountered, that being its inability to locate aircraft or missiles flying at a lower altitude than itself. This is to be remedied as part of a mid-life update programme which will convert the Royal Navy's Sea Harrier FRS Mk 1s into FRS Mk 2s. The main change will be the replacement of the Blue Fox radar with the Ferranti Blue Vixen pulse doppler radar giving a 'look-down, shoot-down' capability. The armament system will be upgraded to use the Hughes AIM-120 Advanced Medium Range Air-to-Air Missile (AMRAAM) although it will still be capable of using the Sidewinder. The FRS Mk 2 will also have the capability to use the Martel or Harpoon air-to-surface missile or the Sea Eagle anti-ship missile. There will also be provision for two podded Aden 30mm cannons to be carried under the fuselage. The opportunity will also be taken to introduce the future standard NATO

The first Sea Harrier FRS Mk 2 ZA195 made its maiden flight from Dunsfold on 19 September 1988. The modified nose is to house the Ferranti Blue Vixen multi-mode fire control radar. (B.Ae.)

Joint Tactical Information Distribution System (JTIDS) of secure voice and data links as well as an improved radar warning system and a new advanced cockpit.

The feasibility study for the Sea Harrier FRS Mk 2 started in 1983 and it is expected that the first converted aircraft will enter service in 1989. As part of the development programme two ex-civil BAe125-600s are being converted for use as trials aircraft. These could possibly be used for training purposes after they have completed their development work, and will possibly be the first BAe125s to fly in Royal Navy markings. The current plans are to convert 33 of the Sea Harrier FRS Mk 1s to FRS Mk 2 standard, but it is possible that the remainder will also be converted and subsequently maybe even some new build aircraft ordered. The Sea Harrier FRS Mk 2 will continue in service well into the 21st Century when it will possibly be replaced by what will be in effect a supersonic Sea Harrier.

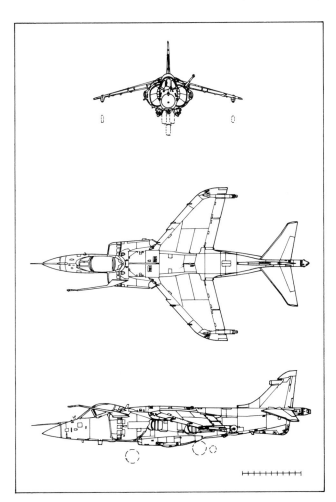

43   Sea Harrier FRS Mk 1

PRODUCTION

FRS Mk 1   XZ438 – XZ440, XZ450 – XZ460, XZ491 – XZ500, ZA174 – ZA177, ZA190 – ZA195, ZD578 – ZD582, ZD607 – ZD615, ZE690 – ZE698.
Harrier T Mk 4N   ZB604 – ZB606.

CHAPTER FORTY-FOUR

Dassault Falcon 20

Dassault started design work on an executive turbojet powered aircraft, known as the Mystere 20, in 1960 with the go ahead for the manufacture of a prototype being given in December 1961. The Mystere 20 was a ten-seat executive aircraft powered by two tail mounted turbojet engines. The first prototype, F-WLKB, powered by two 3000lb thrust Pratt and Whitney JT-12A-8 engines, flew for the first time on 4 May 1963 and a year later the engines were replaced by General Electric CF-700-2B turbo-fans. Dassault were soon very successful in obtaining an order for a large quantity of Mystere 20s for the Business Jet Sales Division of Pan American Airlines and in all over 200 were supplied to the USA

out of some 400 sold world wide. During 1966 the Americans adopted the name "Fan Jet Falcon" for the type and later the name "Falcon 20" was more generally adopted. During 1976 a military version of the Falcon 20, the HU-25A, was sold to the US Coast Guard to meet a requirement for a medium range surveillance aircraft. In addition to the introduction of special mission equipment new Garrett AiResearch ATF3-6-2c turbofan engines were used on this version.

During 1983 the existing contract covering the management and operation of the Fleet Requirement and Aircraft Direction Unit (FRADU) became due for renewal and was offered on competitive tender to suitable companies. The successful bid was received from Flight Refuelling Aviation Ltd (FRA), who took over the operation of FRADU from Airwork Ltd, which had run the Fleet Requirements Units since

Falcon 20, N905FR, operated by Flight Refuelling Aviation on FRADU duties carrying AST-4 pods, which simulate airborne and missile radars. (Flight Refuelling)

January 1950 and the combined FRU and ADTU from December 1972.

Shortly after taking over the unit Flight Refuelling was tasked with finding a replacement for the ageing Canberra T Mk 22s being used by FRADU. The aim was to find a suitable aircraft to replace the Canberra then being used principally in radio and radar calibration roles but in addition for simulating anti-ship missile attacks on Royal Navy ships. The Falcon 20 was selected as the most suitable aircraft available capable of being adapted to carry out the wide range of roles performed by FRADU's Canberras. Initially FRA obtained ten surplus Falcon 20s from the Federal Express Corporation of Memphis, in the USA. These aircraft were Falcon 20Ds which had been retrospectively modified, having had the standard cabin door replaced by a large hydraulically operated freight door 5ft high by 6ft 2in wide.

The Falcon 20s started to arrive at the FRA company headquarters at Hurn towards the end of 1984 and the first aircraft, N901FR, was formally 'presented' to the Royal Navy at a ceremony at RNAS Yeovilton on 5 February 1985, which was attended by Michael Cobham, the chairman of the Flight Refuelling Group, and Rear Admiral L. E. Middleton, Flag Officer Naval Air Command (FONAC). However neither this nor any of the Falcon 20s subsequently delivered have been taken on charge by the Royal Navy, being operated on a charter basis as civil aircraft, finished in the FRA colour scheme and with the first twelve carrying US civil registrations and the remaining four with British civil markings.

To carry out the roles previously performed by Canberras, the Falcon 20s were modified to change the equipment fit and to introduce four hardpoints under the wings to carry the variety of pods necessary to perform the whole range of required activities. Initially the Falcon 20s supplemented and in time replaced FRADU's Canberra T Mk 22s and five Falcons are currently operating in that role. Three Falcons are also used in the electronic warfare role, regularly operating in conjunction with the Canberra T Mk 17s of the joint RAF/RN operated No. 360 Squadron. The remaining two are operating clean, mainly for tracking by radar controllers on board ships, a role previously carried out mainly by FRADU's Hunters. This will probably result in a reduction in the number of Hunters operated by FRADU. However, as the Falcon 20 is not suitable for operations with weapons, it will not totally replace FRADU's Hunters. The introduction of the Falcons has also provided an additional benefit to the unit in that the aircraft are capable of transporting their own ground crew when they go to detachment, something that was not possible with their previous equipment.

During 1986, FRA's contract was extended to provide the complete aerial target services for the Royal Navy for a period of five years. To meet this new requirement FRA increased their Falcon 20 fleet from ten to 16. At the time of writing the Falcon 20 is being developed for the target towing role, two currently being used in this role along with the last remaining two Canberra TT Mk 18s which are also likely to be replaced in the near future.

PRODUCTION

N900FR – N911FR, G-FRAA – G-FRAD.

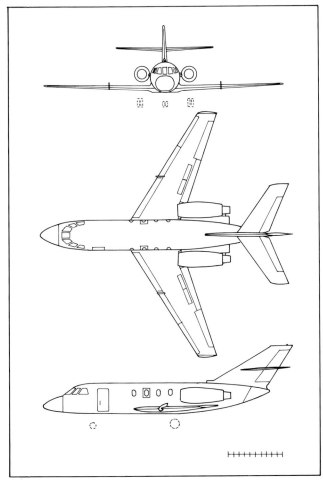

44   Falcon 20